POLYMER SCIENCE AND TECHNOLOGY

NEW DEVELOPMENTS IN POLYLACTIC ACID RESEARCH

POLYMER SCIENCE AND TECHNOLOGY

POLYMER SCIENCE AND TECHNOLOGY

NEW DEVELOPMENTS IN POLYLACTIC ACID RESEARCH

COURTNEY WINTHROP
EDITOR

New York

Additional color graphics may be available in the e-book version of this book.

Library of Congress Cataloging-in-Publication Data

New developments in polylactic acid research / [edited by] Courtney Winthrop.
pages cm. -- (Polymer science and technology)
Includes bibliographical references and index.
ISBN 978-1-63463-054-2 (hardcover)
1. Polylactic acid. 2. Biodegradable plastics. I. Winthrop, Courtney, editor.
TP1180.B55N49 2014
668.4'11--dc23

2014035760

Published by Nova Science Publishers, Inc. † New York

Contents

Preface

Polylactic acid (PLA) is a synthetic and biodegradable polymer. PLA can have important applications in tissue engineering as three-dimensional porous structures. Biodegradable polymers are used in medicine to replace and increase the volume of tissues, as tissue supports and also for controlled drug release. One of the most widely used biodegradable polymers is polylactic acid (PLA), as it is easily biodegraded. This material is used as microspheres or microcapsules to host a broad diversity of drugs, such as antiinflamatory, anticonceptives, narcotics antagonists, local anesthetics and vaccines. This book discusses new developments in the research of polylactic acid.

Chapter 1 - Bioresorbable polymers such as poly-D.L-lactide and poly-lactide-glycolide copolymers have been used as coatings for metallic stents since 2000. They are drug eluting stents (DES). These coatings are impregnated with drugs to deliver anti-inflammatory agents to the arterial lesion site. They have shown to be proven superior to bare metal stents (BMS) by reducing restenosis in coronary stent applications. Many different stent platforms exist and new paradigm of stents, bioresorbable stents (BRS) are currently under clinical evaluation. These stents have different characteristics from stent geometric shape and thickness, design patterns, coating materials, resorption rate, radio-opacity and anti-inflammatory agents. In this chapter, emphasis is driven towards the current stent coating and processing methods of poly-lactide and degradation of poly-lactide for stent applications. This chapter further discusses coating technologies of PLGA and PDLA for drug eluting stents, material designs, degradation of poly-lactide stents, preclinical and initial clinical experimental studies with bioresorbable stents.

Chapter 2 - Biodegradable polymers are used in Medicine to replace and increase the volume of tissues, as tissue supports and also for controlled drug release. The most widely used biodegradable polymers are polylactic acid (PLA), polyglycolic acid (PGA), and ε-caprolactone and their copolymers, as they are easily biodegraded. These materials are used as microspheres or microcapsules to host a broad diversity of drugs, such as antiinflamatory, anticonceptives, narcotics antagonists, local anesthetics and vaccines. Release of drugs from these systems follows different mechanisms, including diffusion and hydrolytic or enzimatic degradation of the polymer. Biodegradability depends on the chemical structure of the material and on the composition of the final product. In many cases a fast degradation is required and no polymer is used; a slower release is observed when using PLA, and the combined use of PLA and PGA copolymers permits to modulate the release degree. The ability of the polymer to control release of active molecules depends on its degradation profile: on controlling the degradation rate of the polymer degradation we can control the release of the hosted active molecule.

Other family of compounds which are currently also being used for controlled release of drugs are layered double hydroxides, LDH, with the hydrotalcite-like structure. These compounds are able to encapsulate the drugs. Their structure consists of stacked layers of $[M^{II}(OH)_6]$ octahedra sharing edges, where a partial M^{II}/M^{III} isomorphic substitution has taken place, and the positive charge in excess is balanced by anions intercalated between the sheets. These anions can be exchanged, introducing anionic drugs, or the solid can be even directly synthetized with the drug within the interlayer space. Formation of LDH/drug systems increases the solubility of the drug, affecting its effect on the living organism and the ability for being transported in the body fluids. The solubility is also related to release, absorption and bioassimilation of the drug.

The nanocompounds which study is here reported can be considered organic-inorganic nanohybrids integrated in polymers forming nanocomposites. LDHs are used to host a given drug and then this organic-inorganic nanohybrid system is supported on a biodegradable polymer, such as PLA, which sometimes induces a partial exfoliation of the LDH sheets. PLA markedly delays release of the drug from several hours when using a LDH-drug system, to several weeks or even months in the presence of PLA. The authors have summarized the current knowledge on these systems, paying special attention to the effect of different preparation procedures on the physicochemical properties (characterization) and release ability.

Chapter 3 - In the last 50 years, it has numerously been evidenced that plasma treatments can effectively change the surface properties of conventional polymers (PP, PET, PE, PA…). Only recently the influence of plasmas on biodegradable polymers, such as polylactic acid (PLA) has been reported. PLA can have important biomedical applications in tissue engineering as three-dimensional porous structures (scaffolds). But their low surface energy leads to poor cell attachment and proliferation and this limits the success of these biodegradable scaffolds. The response it elicits from the surrounding biological environment is crucial and this response is mainly governed by the scaffold surface properties. Therefore, PLA surface properties need to be modified to introduce additional functional groups, which can be recognized as adhesion sites for surrounding cells. Different methods have been developed to obtain the wanted surface properties, however, in the past decade, the use of non-equilibrium plasmas for selective surface modification has been a rapidly growing research field. This chapter therefore presents a critical overview on recent advances in plasma-assisted surface modification of PLA.

Chapter 4 - Polylactic acid (PLA) is well known as a fully biodegradable and sustainable polymer. Due to its high price in the past, its initial applications are mainly biomedical materials, such as bone fixation devices and implant meshes. Due to the increasing awareness of environment protection and sustainable development, the “green” polymer, PLA, has been developing rapidly. The applications of PLA have expanded into plastics, textiles, composites, etc. Since most of the traditional composite materials, such as carbon fibre or glass fibre reinforced epoxy composites, are composed of two distinct fossil fuel based components, they are not eco-friendly and difficult to recycle. Development of biodegradable composites is becoming more attractive. This chapter reviews the development of PLA based biodegradable composites, includes natural fibre reinforced PLA and PLA fibre reinforced composite systems. The main manufacturing processes will be described along with key material and process parameters. The future of these PLA based composites will be explored.

Chapter 5 - 1) Use in Sutures: a) Absorbables sutures with polylactic. Polyglycolic acid, used since 1968, is obtained by polymerization of glycolic acid, the acid which exists in the sugar cane. It is a synthetic product, it is absorbed on 9 months at the most by chemical hydrolysis. It causes minimal tissue reaction and its tensile strength is superior to catgut, silk and cotton. Some derivates are used in Surgery: Polydioxanone acid, glycolic Poly caprolactone, Lactomer Copolymer, Polyglecaprone 25, Polyglactin 910.

b) Self-adhesions sutures. A bearded self-fixing system allowing the suture material for tissue approximation, which eliminates a long and tedious step in wound closure: the knoting to attach and secure the connection. This new system provides a more uniform tension on the soft tissue as they approach so stress is distributed over the entire wound lenght; thus, placement of the sutures with self-locking is less traumatic to tissue as it eliminates the need for sutures or discontinued knots.

c) Active sutures. The new active Plus Vicryl (polyglactin 910 antibacterial coated with triclosan) has been proved as an option for closing wounds, especially in those with active infection or septic environment.

This suture has more effective antibacterial ability, very useful in the prevention of postsurgical infection even using prosthetic material.

Vicryl Plus, the first antibacterial suture was a novel antibacterial designed synthetic suture preventing bacterial colonization in the suture. Vicryl Plus antibacterial agent incorporates Irgacare®, a broad-spectrum antiseptic in form pure triclosan and is known for its effectiveness against bacteria that cause most infections surgical, among which are included Staphylococcus aureus, S. epidermidis sensitive and resistant to methicillin.

Vicryl Plus, to be coated with triclosan, generates a zone of inhibition around the suture and decreases the risk of infection by preventing colonization by bacteria. This eliminates one of the factors contribute to infection, because 66 % of surgical infections are located exclusively on wound.

2) Use in meshes: a) Hibrid synthetic meshes. *Lightweight prostheses:* these meshcs have a big pore and low density. This prostheses *lightweight* can be made so types of material (one single material or mixed materials such non-absorbable and absorbable, such the polylactid acid).

These large pore and low density implants causes less strange body reaction, because have less biomaterial, with the advantage that the implant maintains the elasticity of the host tissue. Lightweight prostheses can be either non-absorbable material as absorbable material, thereby generating hybrid prostheses. These hybrid prostheses generates less fibrous reaction favoring the integration of the mesh and obtaining improved clinical outcomes compared with open surgery.

Composites prostheses: This prosthesis provokes high formation of peritoneum and high integration. The composite prosthesis consists in: Two different- materials layers. The first (superior layer) use to be prosthesis of reticular type, destined to increase the fibroblastic reaction, polypropylene or polyester usually. The second component (inferior layer) uses to be a laminar type and it can be absorbable or scarcely reactive material like the PTFE.

b) Absorbable Synthetic meshes. Up to the present, prostheses used for reconstruction surgery of the abdominal wall have been non-absorbable, which could be associated with chronic inflammatory processes causing complications in these meshes, such as chronic pain, infection, migration, intestinal obstruction, and fistulisation.

New degradable and absorbable materials may be the solution to these complications that, although infrequent, often become quite serious when they appear. These materials may also be applied to potentially contaminated areas. New synthetic absorbable prostheses have appeared recently such materials PGA-TMC (Bioabsorbable Plug Gore®), VIPRO (Vycryl, Ethicon®). These meshes have been used in various studies and they have shown decrease of postoperative pain, which is associated with smaller inflammatory process, being these materials degradable and absorbable.

Other proprieties of these meshes is the hydrophilic characterization, allowing the possibility of incorporates antiseptics or similar in water solution.

c) Absorbable grips in meshes (selfgripping- meshes). This is a partially absorbable mesh, low molecular weight polypropylene monofilament composed of polylactic acid (absorbable). The appendixes of polylactic acid, as a hook, are inserted between the muscle fibers, allowing secure attachment from its inner surface, with less risk of mobilization of the hernia sac under the prosthesis.

It is presented as the first true system of tension-free repair in the treatment of hernia pathology, with extensive coverage of the area being treated. The mesh Parietenne ® Progrip ® (Sofradim, Group Covidien, Trévoux, France) is presented in three modalities, two of them with an elliptical shape, specially designed to fit the anatomy of the inguinal region, with a self-adhesive flap adjustable cord around, either left or right. The third variant is a flat plate 15x9 cm, used for anterior abdominal wall.

The application of polylactic acid in surgery, has developed multiple advances as sutures and surgical mesh. Innovations are not only fit, also in coatings and connections with other products. The future is headed in abdominal wall surgery to mesh with absorbable polyglycolic acid derivatives in the short-medium term, with incorporation of biological products (stem cells, growth factors ...) and tridimensional disposition to facilitate the formation of new tissues the body itself.

Chapter 6 – The authors have been developing biocompatible and biodegradable polymeric micro-nanospheres with poly(D,L-lactide-co-glycolide) (PLGA), loaded with bioactive substance to create new functional medical system toward next generation. It was found that the function of such particulate system was determined by its critical micromeritic property such as particle size, shape and specific surface area and compatibility with biosubstance. The PLGA nanospheres (NS) can improve the drug absorption at the target site due to sustained stay and prolonged release of the drug, leading to the increase in pharmacological effect. It was also found that PLGA NS can be transferred into the cells by endocytosis, especially phagocytosis. This function can be applied to deliver orally nucleic acid substances such as siRNA, asORN and decoy ODN etc. by administrating drug loaded PLGA NS. Further nanomedical device such as eluting stent or balloon was created by depositing PLGA NS loaded with drug on the surface of bared stent or balloon. Although several preparation methods of PLGA NS have been proposed, the authors employed the Emulsion Solvent Diffusion (ESD) method developed by us for industrialization of PLGA NS process due to its simple process requiring no special apparatus and easy scaling up. By using this process many kinds of functional cosmetics and hair restoration tonics containing the PLGA NS have been developed and launched on the market. Several drugs and medical devices using the author's PLGA NS are now under clinical trials managed by pharmaceutical companies and universities.

Chapter 7 - Reactive compatibilization of polylactic acid (PLA)/thermoplastic starch (TPS) blends using methyl methacrylate-glycidyl methacrylate (GMA-MMA) copolymer as compatibilizer was reported in this chapter. During melt blending, the reduction of torque of the uncompatibilized PLA/TPS blends was related to the breakup of starch chains. Moreover, the chemical reaction between MMA-GMA copolymer with PLA and TPS resulted, on one hand, the chain extension of PLA and, on the other hand, torque equilibrium due to the competition among glycerol-MMA-GMA reaction, starch-chain breakup and the formation of the actual compatibilizer MMA-GMA-g-starch. The effect of melt blending time and the concentration of MMA-GMA copolymer on the compatibilization reaction was studied. As the reaction time increased the formation of the actual compatibilizer produced a drop of the TPS domains. On the other hand, the variation of MMA-GMA copolymer concentration evidenced a competition between PLA chain extension and the formation of the actual compatibilizer MMA-GMA-g-starch. Reactive compatibilization of PLA/TPS blends resulted in ductile materials which could have potential application in packaging and agricultural mulch.

In: New Developments in Polylactic Acid Research
Editor: Courtney Winthrop
ISBN: 978-1-63463-054-2

Chapter 1

Drug Eluting Stents and Polylactide Stents

Tré Raymond Welch, PhD*[*1]
and Surendranath R. Veeram Reddy, MD[2]
[1]Department of Cardiovascular and Thoracic Surgery,
University of Texas Southwestern Medical Center, Dallas, Texas
[2]Department of Pediatrics, University of Texas Southwestern Medical Center,
Dallas, Texas, US

Abstract

Bioresorbable polymers such as poly-D.L-lactide and poly-lactide-glycolide copolymers have been used as coatings for metallic stents since 2000. They are drug eluting stents (DES). These coatings are impregnated with drugs to deliver anti-inflammatory agents to the arterial lesion site. They have shown to be proven superior to bare metal stents (BMS) by reducing restenosis in coronary stent applications. Many different stent platforms exist and new paradigm of stents, bioresorbable stents (BRS) are currently under clinical evaluation. These stents have different characteristics from stent geometric shape and thickness, design patterns, coating materials, resorption rate, radio-opacity and anti-inflammatory agents. In this chapter, emphasis is driven towards the current stent coating and processing methods of poly-lactide and degradation of poly-lactide for stent applications. This chapter further discusses coating technologies of PLGA and PDLA for drug eluting stents, material designs, degradation of poly-lactide stents, preclinical and initial clinical experimental studies with bioresorbable stents.

* Corresponding Author: Tré R. Welch, PhD, Instructor at UT Southwestern Medical Center of Dallas, 5323 Harry Hines Blvd. 75390. E-mail: tre.welch@utsouthwestern.edu

Introduction

Use of metal stents has revolutionized the treatment of coronary artery disease and later congenital and structural heart disease. The first generation of stents were bare metal stents (BMS). Metal stents elicit an immunological response of inflammatory mediators leading to neointimal hyperplasia. The chronic inflammatory response elicited by the permanent metal stent leads to a narrowing of the artery called restenosis. The long term disadvantages of metal stents, late in-stent stenosis and thrombosis, has necessitated the need for prolonged antiplatelet therapy. To counteract this problem, drug-eluting metal stents have attracted much attention however due to late thrombosis; interest has been rekindled in bioresorbable stents (BRS).

Stent utilization is widespread and an important tool in the management of congenital heart disease (CHD). Despite advances in engineering, and the proliferation of coronary artery stent development, essentially the same technology is in clinical use for CHD applications today. The most important disadvantage of BMS for pediatric population is the permanent nature of these metal stents, which will not grow with the patient and will impose additional surgical procedures for their removal. Currently there are no BRS in market that are specifically designed for treatment of CHD. There is tremendous need for BRS that will provide the early advantages of a metal stent and will eventually disappear without causing the late disadvantages of in stent stenosis and thrombosis. And most importantly after complete resorption, these temporary BRS implants will allow for normal growth and function of the blood vessel.

Bioresorbable stents were first developed in 1983 at Duke University. The "Duke Stent" was a first in a paradigm of bioresorbable technology that began the investigation of implantable bioresorbable medical devices. The current biodegradable stents are made of either polymers like Poly-L-Lactic Acid (PLLA) or metal alloys like magnesium etc. The PLLA material is the most frequently used polymer to build biodegradable stents. This material already has many human applications and is used in resorbable sutures, soft tissue implants, orthopedic implants and dialysis media. The Igaki-Tamai stent is the first absorbable stent that was implanted in humans [1]. Multiple BDS are undergoing pre-clinical and clinical trials and are commercially available throughout the world [2-4].

Bioresorbable materials used for implantable devices range from polyesters, poly-anhydrides, poly(orthoesters), polyurethanes, poly(ether amides), poly(amino acids) and tyrosine-derived polycarbonates. Polyesters have been used in many medical implantable applications. More commonly used polyesters are Poly-L-lactide (PLLA), poly(D,L-Lactide) (PDLA), polyglycolic acid (PGA), polycaprolactone (PCL), Poly(trimethlyene carbonate). PLLA is composed of the (L) enantiomer of the lactic acid monomer with a homopolymer that is semicrystalline. PDLA copolymer is an amorphous polymer due to the racemic mixture of monomer that disrupts the crystallinity. Since it is amorphous, it degrades much faster than PLLA that usually degrades over several years. Copolymers of PGA and PLA have been synthesized to formulate poly(L-lactide co-glycolide) in different ratios the tailor the degradation of the polymer from a few weeks to a few months with PGA usually degrading between 6-12 months and are discussed as coating for stents.

Coating Methods for Stents

Dip Coating

Stent dip coating has been performed using a single drug polymer solution [5, 6]. This simple method uses a drug prepared in an organic solvent such as tetrahydrofuran (THF) that is miscible with Poly-lactide co-glycolide (PLGA) at a certain concentration. The uniqueness of using biodegradable polymers is they adhere to metal stent surfaces and control the drug release. The stent design can be modified to have pores, channels, or texturized surfaces. To create surface modifications they are dipped into the organic solution for a period of time to permit the solution to permeate into the stent pores or channels [7]. The concentration of the drug into the polymer can be increased to reach a therapeutic level relevant to the patient. Once the coating has dried and the solvent evaporates from the surface, another dip into the solution can be performed to create another coating layer on the stent surface. The last dip coating solution can be the polymer without drug to be used as a diffusion barrier to control the drug release from the stent surface. The limitation of this method requires the use of multiple dips into the polymer solution to create higher concentrations of the drug. The drug to polymer ratio can only saturate the solution to a limit thus creating equilibrium between the solution concentration and the coating amount. These dip coatings have a typical thickness range from 10-80 μm [5, 8-10].

A layer coating on PLLA has been investigated by Wang et al, [5] shown in Figure 1. They used a cast bi-layer of PLGA with sirolimus on top of a layer of PLLA to simulate drug release from a PLLA substrate. They also created a tri-layer of PLGA 50/50 on top of the PLGA 53/47 with sirolimus on top of a base layer of PLLA to simulate drug release from a fully biodegradable multilayer stent. Their results show the drug release is biphasic with intial diffusion control stage followed by a degradation-controlled stage. The amount of drug loading had no effect on these drug release kinetics. The use of the tri-layer coating retarded the release primarily due to the faster water uptake into the 50/50 layer and the partitioning of the sirolimus into the 50/50 layer.

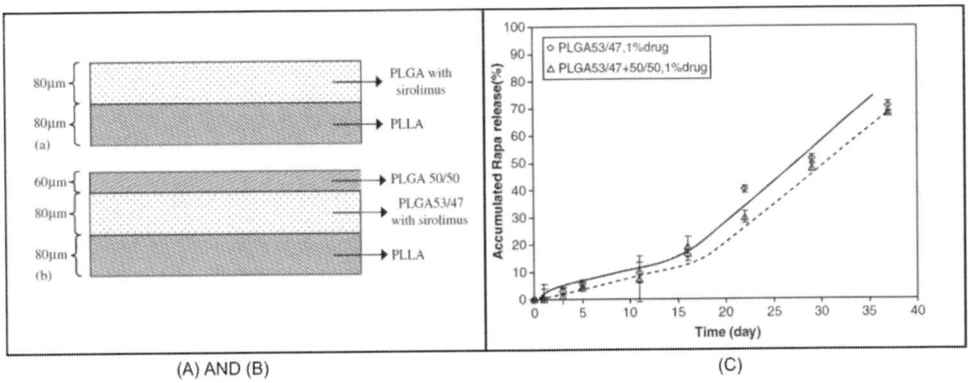

Figure 1. (A) Showing a diagram of the bi-layered polymer matrix; (B) a tri-layer matrix with respective thickness;(C) is the cumulative drug release from the bi-layered and tri-layered film. This article was published in Biomaterials, Wang X, Venkatraman SS, Boey FY, Loo JS, Tan LP, vol. 27, Controlled release of sirolimus from a multilayered PLGA stent matrix, pp. 5588-95. Copyright Elsevier (2006), with permission from Elsevier.

An alternative method for dip coating is spin coating a stent. This method uses a stent that is mounted on a mandrel, and the stent is rotated at a high or low speed. The stent is placed tangent to the surface of a solution bath or the solution is dropped onto the stent and then dried and repeated. Spin coating is extremely fast but is limited to using only one substrate at a time. It lacks material efficiency, repeatability, and thin film deformity. Advantages to dip and spin coating deposition methods are these processes are less complex and provide a more uniform coating. The dip process is suitable for coating complex structures and has a high deposition rate. They do not provide for spatial control and is inflexible in terms of switching coating materials.

Polymer spin/dip coating methods have been used to coat they Cypher™ stent by Cordis Corporation (Miami, US). This was the first drug eluting stent to obtain USFDA approval for coronary stent lesions. The sizing of the stent was between 2.5 and 3.5 mm diameter and less than 30 mm length [8, 9]. The delivery system was composed of a Bx velocity balloon expandable coronary stent premounted on a Raptor rail rapid exchange. The drug coating consists of sirolimus of 70-314 μg with a base coat of parylene C and a second coat of polyethylene-co-vinyl-acetate (PEVA) and a final coating of poly-n-butyl methacrylate (PBMA). Parylene C is applied by vapor deposition methods using high temperature and low pressure. This method has also been referred to as conformed coating of stents. Parylene C has been approved by FDA for medical implants showing excellent biostability and biocompatibility. The first coating layers of the CYPHER™ stent (Cordis) are paralyene C, and then drug mixed with PEVA and PBMA and applied to form a second layer. The final layer is PBMA without drug to provide a slow drug release. The CYPHER™ stent was evaluated in the RAVEL trial showing superior antirestenotic effects compared to BMS [11]. The Endeavor®RESOLUTE stent (Medtronic) is an FDA approved zotarolimus-eluting stent. This is a metal stent that uses a dip method to apply phosphorylcholine polymer as a drug carrier. This is not a biodegradable polymer but is a biostatic and biocompatible natural occurring phospholipids polymer. Their clinical results have shown to be effective in treating coronary artery diseases [12-14].

Inkjet Coatings

Inkjet coatings offer another method of drug delivery application using biodegradable polymers such as PLGA and PLA. Lee et al., [15] successfully created and printed microscale particles from paclitaxel-loaded PLGA inks. As noted earlier bare metal stents are coated with an anti-proliferative drug embedded within a polymer to control the drug release to the lesion site. Inkjet dispensing has been reported use in nanofabrication applications including microfluidic systems, drug-eluting stents, photovoltaic arrays, tissue engineering [16, 17]. These applications require quantitative and spatial accuracy for surface deposition to ensure repeatability of feature size, functionality, and analytical accuracy. Figure 2 shows an injet printer from microdrop Technologies GmbH, Germany and a metal stent to be coated with the dispension head on an inkjet printer.

This technology can be subdivided into two catagories: continuous mode and drop on demand inkjet printer. The printer is composed of a driver, transducer, fluid pump, charge electrode, a catcher and a substrate to coat. In continuous mode the pressurized fluid is forced through 50-80 μm diameter orfice forming a liquid jet operating at rates of 1 MHz. Applying a single frequency disturbance to a piezoelectric transducer in the frequency range of the jet causes an amplified disturbance and forms droplets. The charge electrode located near the orfice charges the droplets. If the high voltage plates are misaligned, the charged droplets can land on a catcher cup and return to the liquid reservoir (see Figure 3). Drop on demand (DOD) inkjet printers have the fluid at ambient pressure, and a transducer is used to create the drop when specified by the user at rates of 1 Hz to 25 KHz. Pressure waves are created by the transducer that travel to the orfice and convert this energy into a fluid velocity resulting in a droplet being ejected from orfice.

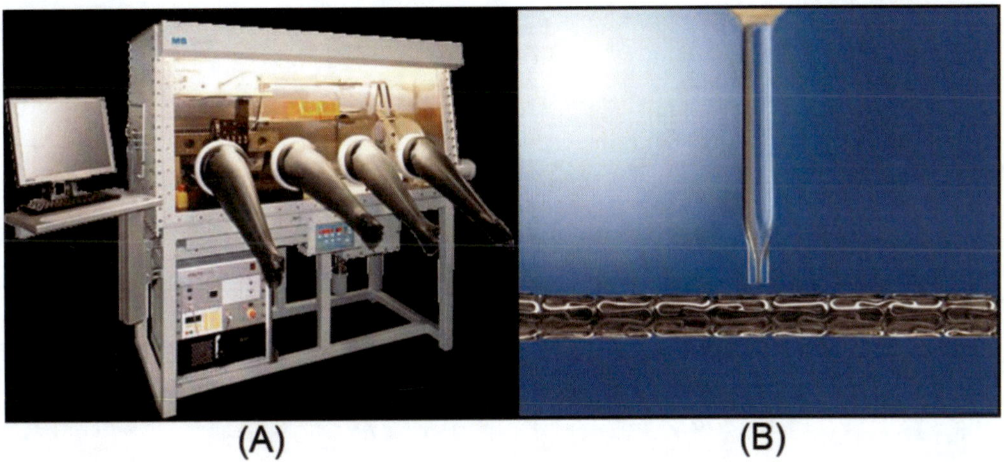

Figure 2. Illustrating (A) Microdispensing Autodrop Platform in Glovebox and (B) a stent coating platform with permission of *microdrop Technologies GmbH, Germany*.

Limitations of the drop-on-demand inkjet printer droplets can be significantly different in mass, velocity, and trajectory from subsequent droplets emerging during establishment of steady-state ejection conditions [16]. DOD inkjet dispensers operate by use of a piezoelectric deformation, electrostatic pull or thermal transduction and the dynamics of the liquid jets and microdroplet generation. The diameters of the droplets are roughly equivalent to the diameters of the orfices from their nozzles [16, 18]. Some of these variablities can result from the driving waveform, the droplet ejection frequency, fluidic pressure in the fluid cavity, orifice surface condition, and the fluid rheological properties. The piezoelectric transducer can cause potential particle entrapment and lead to growth of bubbles within the acoustic cavity.

Newer advances with inkjet coating systems has shown to print on 3D compelx surfaces. In 2014 Perkins et al., [19] have developed a direct-write deposition on 3D complex surfaces. They were able to use a customized direct-write inkjet system developed at MicroFab Technologies, Plano, Tx to deposit polyester urethane urea (PEUU) with paclitaxel drug homogenous layers of onto a titanium surface. The system consists of a pneumatic controller, a JetDrive III external waveform generator; JetLab® software with waveform amplifier; horizontal and vertical plane optic system, and a piezoelectric nozzle with a 50-μm orfice. The JetLab® interface with the appropriate pulse waveform for consistent droplet formation was characterized by manipulating the rise time, dwell time, fall time, voltage and pressure. The motion controller compartment and pattern monitor allowed for precise control of the length, width, and thickness of the coating by adjusting the step size and pitch between the droplets. The step is the distance between the droplets and the pitch is the diameter of the droplet. These parameters were optimized to obtain an evenly distributed coating layer. A CCD camera was attached to the system to ensure consistency of the drop formulation during the coating process.

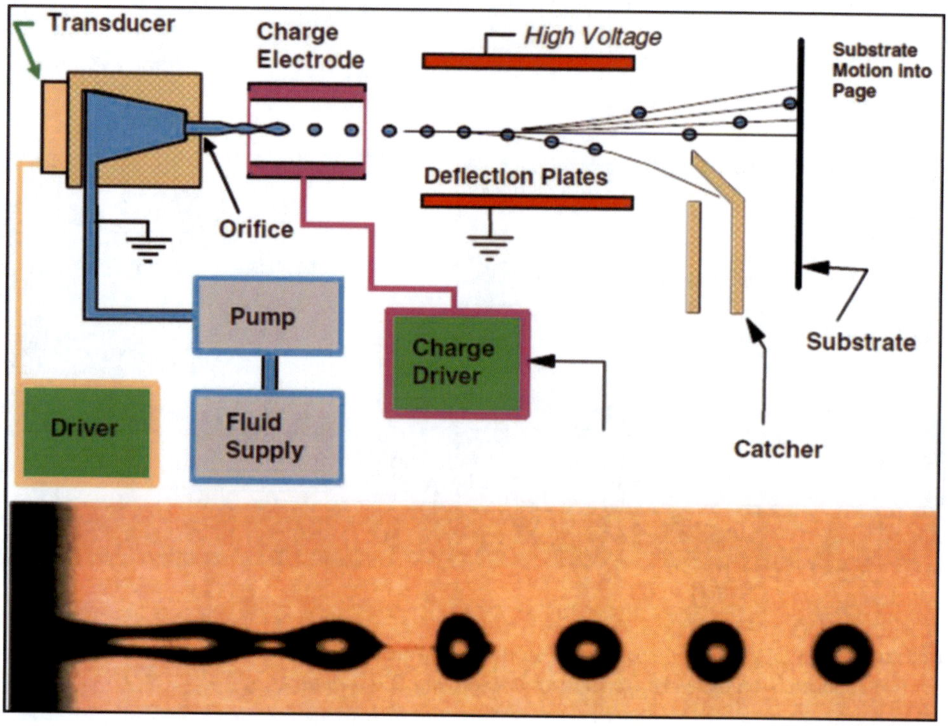

Figure 3. A schematic diagram of a continuous mode inkjet printer. Reprinted from Annals of Biomed Engineering,35, 2007,Tarcha PJ, Verlee D, Hui HW, Setesak J, Antohe B, Radulescu D, Wallace D, The Application of Ink-Jet Technology for the Coating and Loading of Drug-Eluting Stents, Figure 2,1792.Copyright 2007 with kind permission from Springer Science and Business Media.

Their results showed films with less than 10 coating layers had low integrity and adhesion properties to a titanium alloy substrate. While films, with more than 25 coating layers had longer release kinetics with their respective drug loading amounts. They show the choice of drug and number of layers has a significant effect on the drug release kinetics, platelet adhesion, and smooth muscle cell proliferation. They were able to accommodate gradient patterns as well. Using an imaging processing algorithm in Matlab, they were able to print different layers and gradients from a black and white image of a tissue cross-section. This was performed by varying the fluid properties and deposition parameters to generate complex patterns of biomaterials.

Spray Coating

A typical spray coating system uses a spray nozzle and a pump that moves the polymer coating material from the reservoir to the nozzle illustrated in Figure 4 [20-22]. The ejected material is sprayed onto a stent surface as a microdroplet. THF has been used as an organic solvent to dissolve PLGA. Once dissolved, this polymer can be sprayed onto the stent surface. The stent is placed onto a rotating mandrel and the polymer solution is dissolved with a drug such as paclitaxel and connected to an airbrush pistol. The pressurized air pushes through the nozzle aspirating the polymer solution.

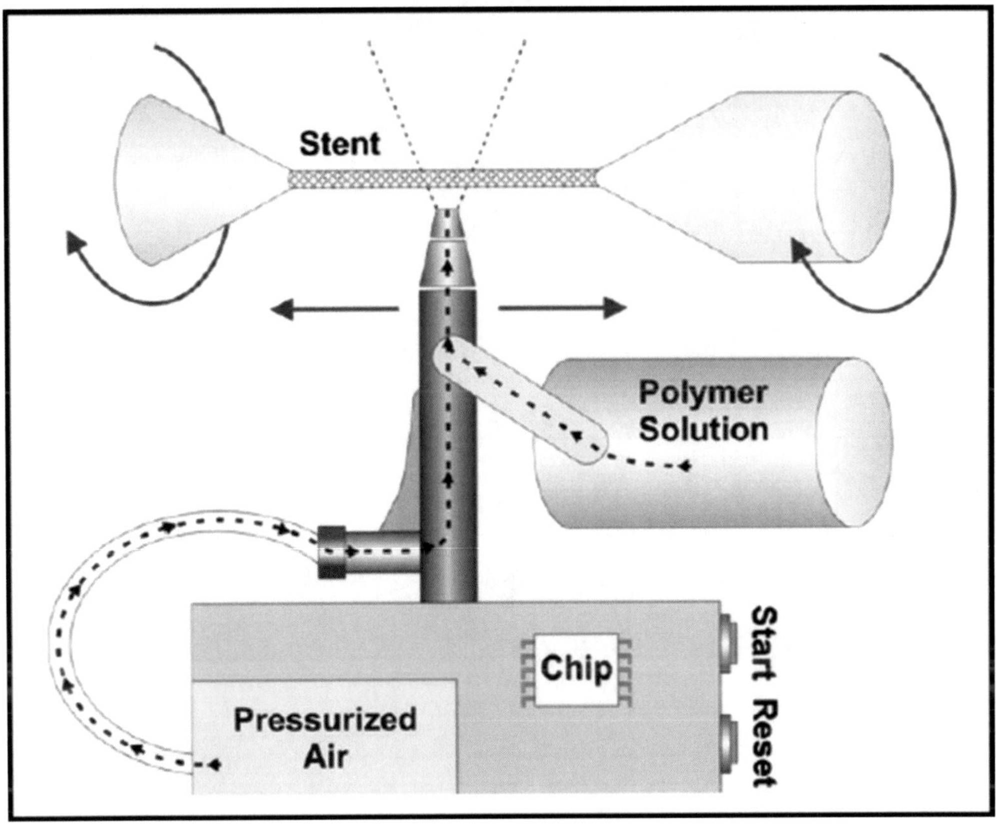

Figure 4. Showing a spray coating apparatus. A horizontal rotating coronary stent is spray-coated with polymer solution consisting of 0.8% Poly(ethylene carbonate) in dichloromethane loaded with Paclitaxel, radiolabelled Paclitaxel, or Coumarin-6. This article was published in European journal of pharmaceutics and biopharmaceutics, Bege N, Steinmuller SO, Kalinowski M, Reul R, Klaus S, Petersen H, Curdy C, Janek J, Kissel T, Vol. 80, Drug eluting stents based on Poly(ethylene carbonate): Optimization of the stent coating process, pp. 562-70. Copyright Elseveir (2012), with permission from Elsevier.

To insure a uniform coating, several passes across the stent and surface until a desired coating is achieved. The stent is cured in a pre-heated oven for three hours at 65-70°C to flash off any excess solvent from the coating.

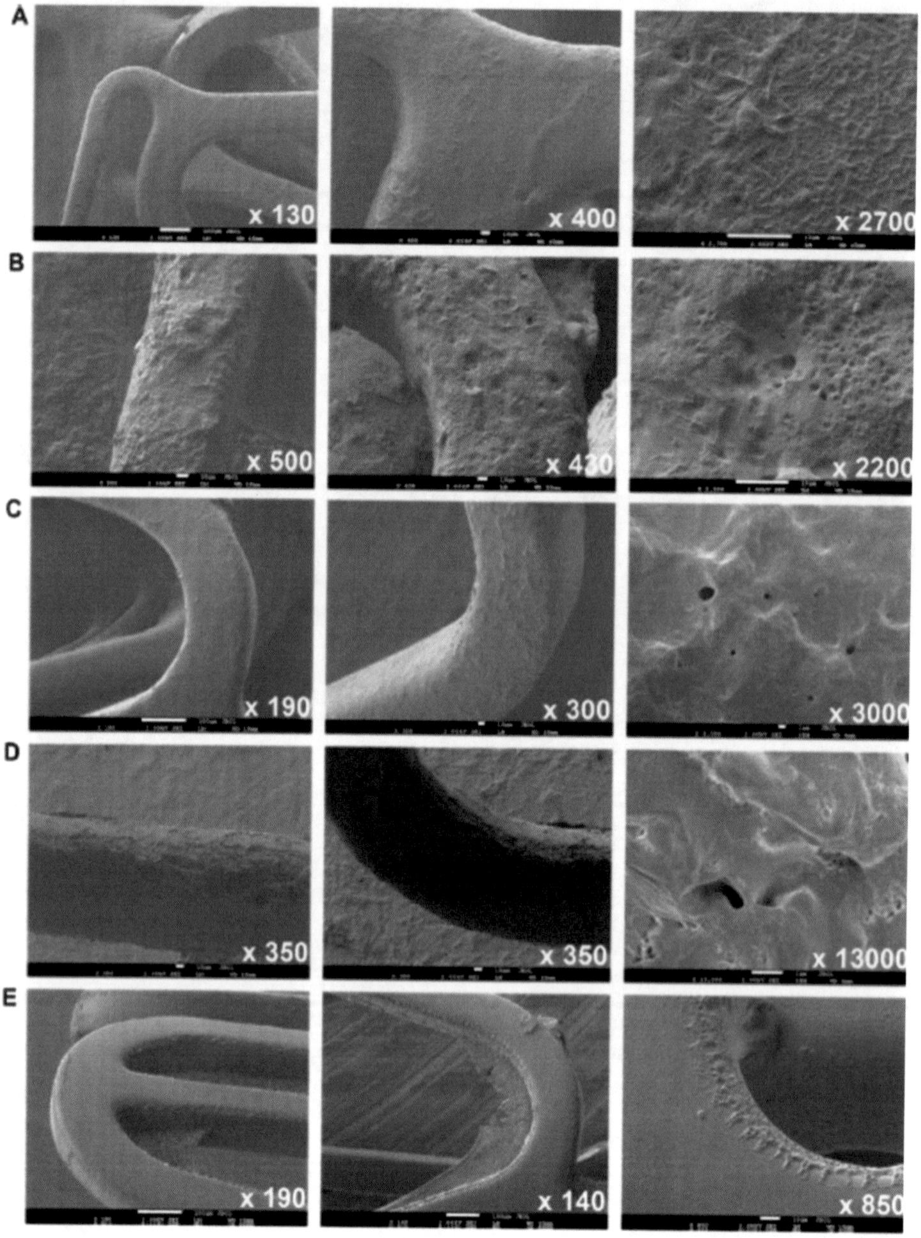

Figure 5. Morphology of coated metal stents. (A) Sprayed stents once with 8% Paclitaxel loaded PEC solution; (B) stents spray coated 3 times with 8% loaded PEC solution; (C) stents spray coated once with 25% Paclitaxel loaded PEC solution; (D) stents spray coated 3 times with 25% Paclitaxel loaded PEC solution; (E) stents dip coated once with 25% Paclitaxel loaded PEC solution. This article was published in European journal of pharmaceutics and biopharmaceutics, Bege N, Steinmuller SO, Kalinowski M, Reul R, Klaus S, Petersen H, Curdy C, Janek J, Kissel T, Vol. 80, Drug eluting stents based on Poly(ethylene carbonate): Optimization of the stent coating process, pp. 562-70. Copyright Elsevier (2012), with permission from Elsevier.

A spray coating method is used on the TAXUS® Express paclitaxel elevating stent (Boston Scientific, Natick, MA). Their coating consists of a mixture of placitaxel and an elastomeric tri-block copolymer poly (styrene-b-isobutylene-b-styrene) (SIBS, that has the trade name Translute) [8, 23-25]. The stent use a stainless steel platform and the drug is released from Translute polymer matrix into the arterial walls without going through the rate-controlling membrane, providing a diffusion control matrix. The in-vitro release studies have varying patterns depending on the amount of placitaxel [24]. The TAXUS® stent was the 2nd FDA approved drug eluting stent in 2004 using 50-200μg of drug [26]. The uniformity of the stent coatings after stent expansion was investigated by Bege et al., [21]. They used polyethylene carbonate (PEC) mixed with paclitaxel, spray coated stents and compared to a dip coated stent. Their results have shown the spray coating is homogenous with uniform roughness with one spray coating (see Figure 5). The more coatings they applied to the stent resulted in a rougher and unevenness of the coating. In comparison to a dip-coated stent in Figure 5E, the coating was less homogenous and the polymer accumulated at the stent bending regions that leads to bridging of the stent.

Spray coating of stents has shown reduced restenosis rates compared to BMS [23, 27-29]. There are limitations to this type of coating such as webbing between struts. Advancement in spray coating has been progressed by Biosensor International Ltd. to produce more degradable coating with less imperfections and overcoating. They incorporated a computer controlled imaging system into their spray coating system. Their system optimizes the coating by determining the transverse path by which the dispenser head travels over the stent and calculates the speed and positional displacements to provide a uniform coating. The converted greyscale image controls the different parameters such as the motion speed, dispensing flowrate and a number of passes to form a more uniform coating. This manipulation allows the control to coat the wider and thinner struts. Also, the distance of travel from the nozzle head to the stent can be adjusted to affect the coating deposition. Their system allows for the coating of thinner parts and along curved features of the stent. The Biomatrix Flex ™ Biosensor International is a stainless steel stent platform coated with abluminal biodegradable polylactid acid (PLA) and Biolinus AG™ [30]. They show an improved flexibility, trackability, and layer initial cell opening. Their design consists of curved connectors that increase flexibility of the stent. This coating is absorbed within 6-9 months as the polymer degrades to lactic acid [31, 32]. Bioimus is more lipophilic than sirolimus with faster drug uptake with excellent immunosuppressant and anti-inflammatory properties.

Ultrasonic Spray Coating

Atomize spray coating machines have been used to coat stents with a uniform and continuous coating. One company, Sono-Tek Corp. (New York, US) uses this platform to coat stents.

The basic principal behind the ultrasonic coating platform is the use of a high frequency ultrasonic atomizing nozzle and a low pressure gas stream to shape the low velocity drops into a narrow soft spray usually operated at a frequency of 120 KHz [33]. The operational requirements use a flow rate of 20-100 μl/min and spray droplet diameters of 0.5-2 mm. Similar to spray coating, the stent is placed on a mandrel attached to a rotating shaft that translates to coat the entire stent length that requires several transverses. The mixture liquid is the polymer and drug dissolved in an organic solvent, i.e. THF, and diluted by 0.5-2%. The use of high vapor pressure solvent causes rapid drying of the coating. The coating of the stents is performed under a nitrogen atmosphere to lower the surface tension of the liquid once it contacts the stent surface.

The driving principal for this coating method is the atomization based upon vibration frequency as shown in Figure 6. Atomization occurs when the liquid solution is placed upon the smooth surface that is vibrating in a motion perpendicular to the surface. The liquid absorbs some of the vibrational energy that is transformed into capillary waves. Upon increasing the amplitude, a critical amplitude is reached. The waves then collapse and liquid droplets are ejected from the tops of the degenerating waves normal to the atomizing surface. The number median droplet diameter, $D_{N,0.5}$, has been determined mathematically by Rayleigh as being inversely proportional to the vibration frequency (*f*) to the 1/3 power with σ as the liquid surface tension and ρ as the liquid density and then modified by others to develop the following relation [34, 35].

$$D_{0.5} = 0.34\left(\frac{8\pi\sigma}{\rho f^2}\right)^{1/3} \qquad \text{Eq.1}$$

The number median droplet diameter size is based upon the frequency of the given nozzle. Using this vibrational frequency for coating metals or polymers breaks up the liquid suspension into droplets thus removing any aggregation of particles and maintaining homogeneity of the suspension mixture. This improves the uniformity and functionality of the surface.

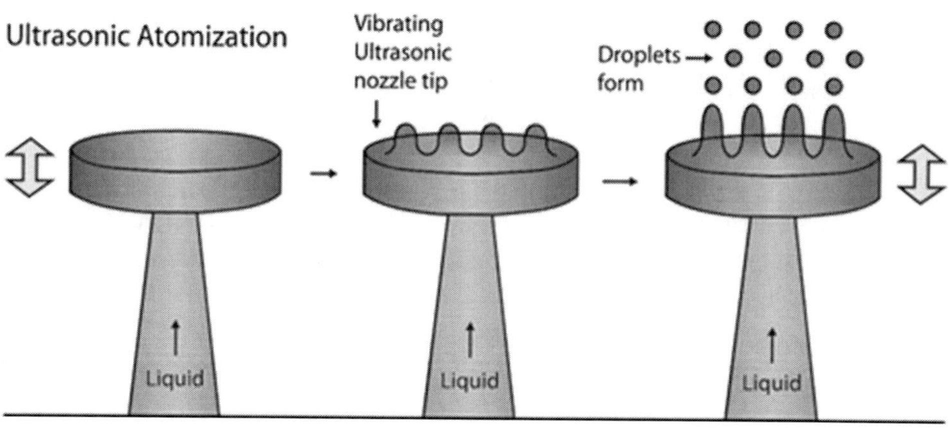

Figure 6. Illustrates the ultrasonic atomization process of and injected liquid being atomized into droplet from the nozzle tip. Image courtesy Sono-Tek Corporation.

Ultrasonic Nozzle Designs

The use of these nozzles has been under investigation for many years. Using the AccuMist nozzle design, compressed gas is forced into a diffusion chamber that produces uniformily distributed airflow around the nozzle stem [33]. The spray at the tip of the nozzle is transported into the gas stream. The spray envelope is conical with a slight bow or hourglass shape and is adjusted by a focusing mechanism to optimize a spray pattern. The distance can affect the spray field of the nozzle. The narrowest beam diameter is at a 1.75 mm focal plane and the median is 13-70 μm.

Using the MicroMist Nozzle (see Figure 7), liquid is injected to the surface through a hypotube [33]. Gas is forced through the nozzle orfice and the gas stream creates a narrow spray of 0.5 mm diameter.

This design uses an independent liquid feed that results in a high degree of spray stability and a greater degree of reproducibility from one spray cycle. The size of the microdroplets depends on the operating frequency ranging from 50-60 Hz. They have a tight size distribution that permits better conformal coating of a surface. Use of these ultrasonic nozzles permits the user to have independent control of the flow rate, spray velocity and drop size. For instance, low flow rates have a high degree of repeatability without compromising spray performance. This maximizes the effectiveness and quality of the coating production.

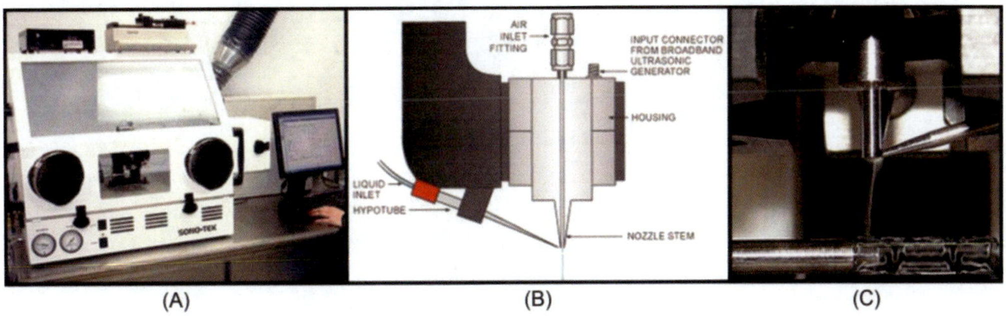

Figure 7. (A) The Sonot-Tek MediCoat DES1000 with a MicroMist nozzle used to coat stents; (B) shows a diagram of the MicroMist nozzle using a hypotube to inject the polymer solution, and (C) is showing a metal stent being coated with a polymer solution. Image courtesy Sono-Tek Corporation.

Morphology of Spray and Ultrasonic Surface Coatings

In practical applications of surface coating the spray liquid is a mixture of polymer and drug dissolved in an organic solvent such as THF, acetone or chloroform. The mixture is usually measured on High Performance Liquid Chromatography (HPLC) for drug quantification. Usually high vapor pressure solvents are used for rapid drying of the droplets on the surface. Characteristics of the coatings can be determined by varying different parameters of the spraying process such as rotational speed, distance from the target surface to the nozzle, flow rate, and the number of transverse passes of the nozzle over the target surface. In Figure 8, the effect of bridging and webbing between the stent struts are due to ineffective optimization of spray parameters.

Investigations into drug dosage with PLGA and varying different system parameters to optimize the stent coatings have been performed by others [7, 36-40]. In 2006, Pan et al., [36] investigated coating 316L stainless steel with a solution mixture of PLGA and an anti-inflammatory drug. They used curcumin and PLGA using three different doses onto a 316L stainless steel stent. Coating with PLGA and no drug showed a coating that was smooth and uniform without cracking or delamination during the stent expansion process and showed no webbing between the struts that have been shown in dip coating methods (see Figure 9). Loading curcumin into PLGA and applying the coating to the stent resulted in a higher surface roughness of the coating without observing any curcumin drug particles on the surface by SEM and AFM measurements. The low dose of curcumin released about 70% in 18 days while a high dosage released only 30% in 18 days. They also investigated use of rapamycin mixed into PLGA and ultrasonic atomized spray coated stents and showed no webbing of the coating between struts and a fine coating layer [38].

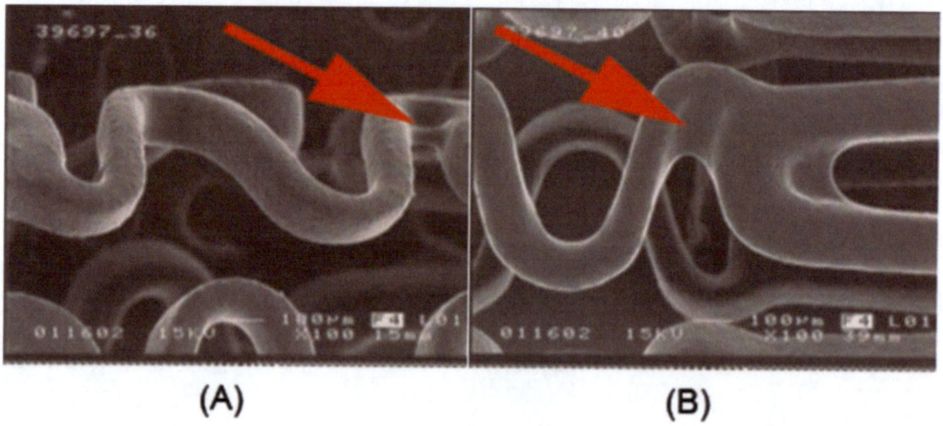

Figure 8. After coating a metal stent without optimized spray parameters (A) shows bridging and (B) shows webbing that can occur. Image courtesy Sono-Tek Corporation.

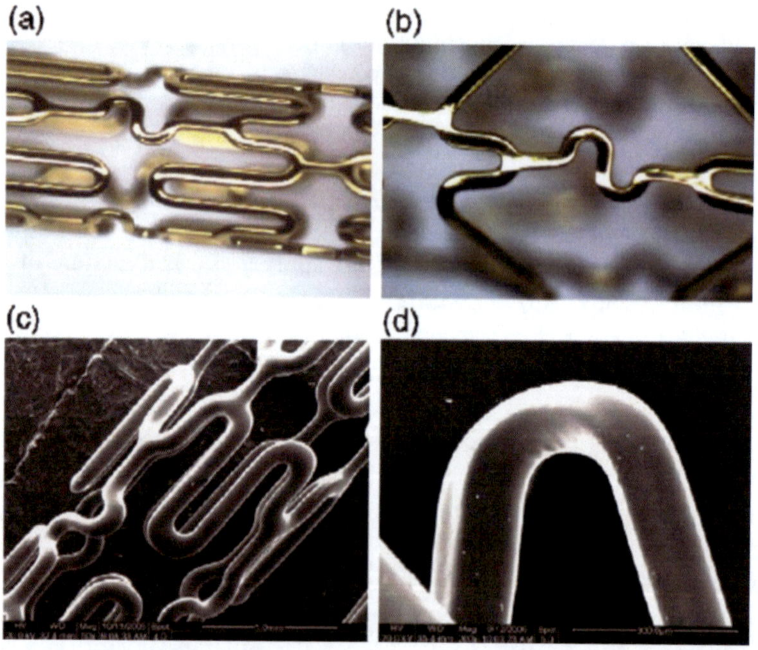

Figure 9. A metal stent ultrasonic spray coated with PLGA and curcumin: (a) 40x optical image before expansion and (b) after expansion; (c) 50x magnification nonexpanded stent and (d) 200x magnification post dilated stent. This article was published in Journal of Controlled Release, Pan CJ, Tan JJ, Weng YJ, Wang J, Huang N, Vol.116, Preparation, characterization and anticoagulation of curcumin-eluting controlled biodegradable coating stents, pp. 42-49. Copyright Elservier (2006), with permission from Elsevier.

Shanshan et al., [37] coated stainless steel stents with PLGA and sirolimus mixture and varied different parameters (solvent, rotational speed, nozzle power, solution flow rate) to evaluate the morphology of the stent coated surface. By varying different solvents from tricholoromethane, dichloromethane, acetone and ethyl acetate, they show the use of acetone resulted in a surface with hollowness attributed due to the low polarity and lower boiling point of acetone. The other solvents resulted in a smooth and uniform surface. The effect of rotational speed of the stent on a rotating mandrel affected the coating. Using 250 rpm the external and internal stent surfaces were uniform and smooth. At 275 and 200 rpm the surfaces were uneven and some scallops formed on the external surface. At 300 rpm the internal surface had more scallop formation but was smooth on the external surface. The use of an optimized rotation speed can make the solution collect on the internal and external surface equally due to the centrifugal force acting on the surfaces of the stent. The effect of nozzle power (Po) was increased by 0.5, P+ 0.5, resulting in a uniform and smooth surface morphology. Further increase of the power resulted in more uneven surfaces due to the formation of smaller liquid drops leading to an increase in specific surface area and faster volatilization rate making the solution more difficult to spread across the stent surface. Reducing the power below Po resulted with no fine drops. Another parameter of interest as mentioned above is the solution flow rate. Using a faster flow rate of 0.07 mL/min increased the coating thickness and was more brittle and wrinkled. At a slow flow rate, 0.03 mL/min small scallops formed on the surface and at 0.05 mL/min showed a uniform a smooth surface. Herein using these parameters for ultrasonic spray coating has led to a more uniform coated stent with PLGA.

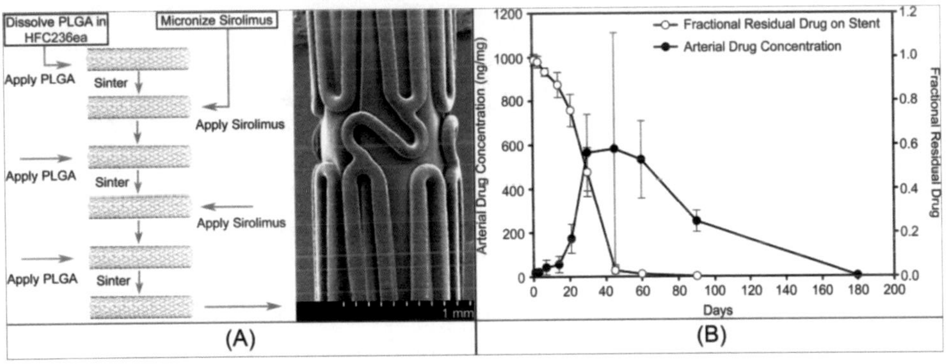

Figure 10. (A) Showing the dry powder deposition process of PLGA and micronized sirolimus by sintering that fuses the layers into a conformal coating. (B) This shows their in-vivo drug release being completed within 45-60 days into porcine coronary arteries over 180 days. The average daily release rate was approximately 3 μg of sirolimus with the data represented as ± standard error. This article was published in Journal of Controlled Release, Carlyle WC, McClain JB, Tzafriri AR, Bailey L, Zani BG, Markham PM, Stanley JRL, Edelman ER, Vol.162, Enhanced drug delivery capabilities from stents coated with absorbable polymer and crystalline drug, pp. 561-67. Copyright Elservier (2012), with permission from Elsevier.

Although these methods presented above are great for stents, a new method using dry powder deposition with PLGA has been investigated [41]. Unlike conventional dip and spray coating methods, this method uses a charge differential to add multiple layers of PLGA and drug shown in Figure 10. In short, PLGA is dissolved into a solvent, HFC23ea, and is pressurized to 4500 psig and sprayed through a nozzle at 110°C. The solvent flashes off coming out the nozzle and the polymer is deposited onto the stent by a charge differential. The stents are sintered to 100°C. The same procedure is repeated with a drug, sirolimus, and deposited to the stent surface. Mutiple layers are formed to create this layered coating on the stent.

Their stent coating showed complete drug release within 45-60 days after stent implantation in Figure 10B. The inclusion of the crystalline drug than the amorphous drug coating resulted with a slow consistent drug release kinetics over a 90 day period and an inflammatory response showing coating loss and neointimal formation in Figure 11.

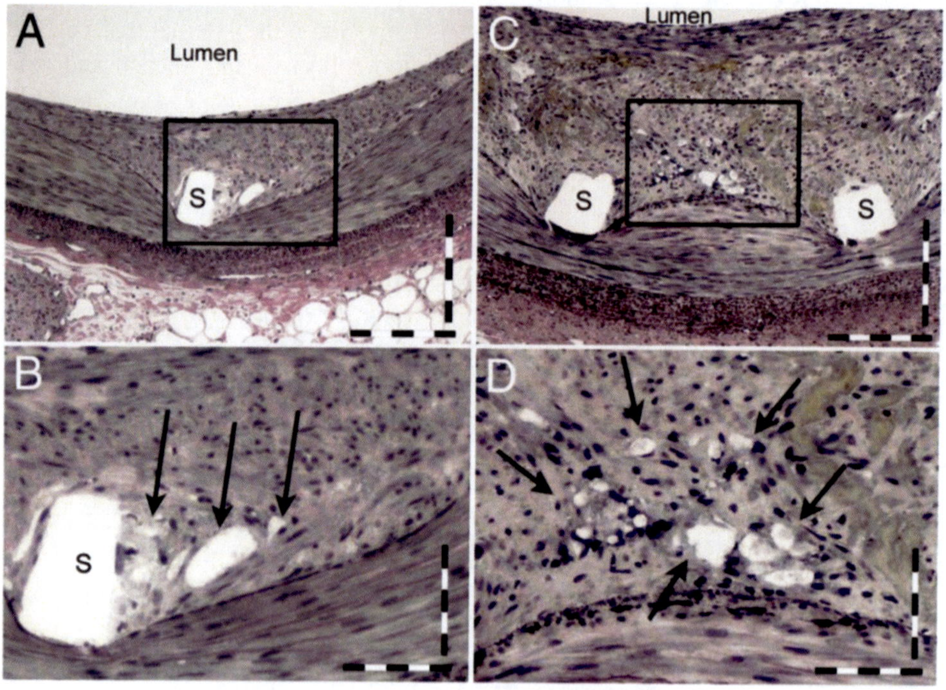

Figure 11. This shows the histological staining of a porcine coronary implant model. The square shaped box is the strut (S). The black arrows are areas of the coating that has been lost during the process. The coating spread into the neointima.The scale bar represents 200 μm (A, C) and 75 μm (B, D).This article was published in Journal of Controlled Release, Carlyle WC, McClain JB, Tzafriri AR, Bailey L, Zani BG, Markham PM, Stanley JRL, Edelman ER, Vol.162, Enhanced drug delivery capabilities from stents coated with absorbable polymer and crystalline drug, pp. 561-67. Copyright Elsevier (2012), with permission from Elsevier.

Degradation of Polylactides

For the polylactide family of polymers, molecular weight degradation occurs primarily by hydrolysis. However, many studies on thermal degradation of polymers has been researched and focused on the reduction in mass and molecular weight of the polymer. In general chemical reactions lead to the changes in polymer material properties associated with degradation through three major pathways [42].

First side group elimination occurs in which the side group is removed from the polymer chain resulting in an unstable polymer accessible that leads to further degradation. The second pathway is random chain scissioning that is similar to hydrolytic chain scission that forms a free radical along the backbone of the polymer chain splitting the polymer chain into lower molecular weight. The third pathway is producing constituent monomer by a free radical mechanism to reverse the polymerization process. These polymers degrade by hydrolysis producing by-products that readily diffuse out of the polymer thus being controlled by bulk degradation kinetics.

Degradation Rate Kinetics

Degradation kinetics of PLLA has been used to predict the hydrolysis rate that depends on the concentration of ester bonds, water, and carboxylic acid and groups [43, 44]. The carboxylic acid and groups are formed by hydrolysis reaction. This is called the autocatalytic model for aliphatic polyester given in Eq. 2.

$$\frac{d[E]}{dt} = -k[COOH][H_2O][E] \qquad \text{Eq.2}$$

with [E], [COOH] and [H_2O] are the concentrations of ester bonds, carboxylic acid and groups and water, respectively and k is the hydrolytic degradation rate constant. This has been further examined by Weir et al., [45, 46] assuming [E] and [H_2O] are constant throughout the degradation process and that [COOH] is inversely proportional to the number average molecular weight (Mn) of the polymer. This has been written as an exponential relationship. Mn_t is the number average molecular weight after degradation time, t, and Mn_0 is the number average molecular weight at t = 0. These assumptions are reasonable if mass loss does not occur because mass loss loss affects the [H_2O] and [COOH] groups. This equation has further been rewritten as

$$\ln Mn_t = -kt + \ln Mn_0 \qquad \text{Eq.3}$$

Using this approach by plotting on a log-linear plot, the hydrolytic degradation rate can be derived from the slopes of the line of the connecting points. This relationship has further been investigated and successfully used with the Arrhenius relationship to predict the acceleration in degradation by increasing temperature. This approach shows the physical degradation of the polymer is a result from a rate constant, k, that is proportional to exp(-Ea/RT). Ea is the Arrhenius activation energy, J/mol; R is the gas constant 8.314, J/mol; and T is the absolute temperature in Kelvins, K, with A as a constant. This method predicts a linear relationship exists between the logarithms of the rate constant versus the reciprocal of temperature given in Eq. 4.

$$\ln k = -\left[\frac{Ea}{RT}\right] + \ln A \qquad \text{Eq.4}$$

In-Vivo Degradation

PLLA is a semi-crystalline polymer composed of crystalline component and random polymer chains that form the amorphous segments. PLLA is composed of crystal lamella interconnected by amorphous tie chains that bind the crystalline structures [47]. These amorphous regions are less packed and more accessible to hydrolytic attack than the semi-crystalline regions of the polymer. Once PLLA is implanted into a hydrated environment such as a phosphate buffered solution, deionized water, or an animal, PLLA materials undergoes different stages of degradation. The first stage is hydration of PLLA. After implanting the medical device, the polymer begins to absorb water from surrounding tissue. The chain end groups are hydrophilic due to the carboxylic acid end groups and they are oriented to the amorphous phase of the polymer. The second stage is depolymerization by hydrolysis that is shown by a reduction in the molecular weight of the polymer that is due to the water attacking the amorphous phase. This causes the mass loss of the polymer that result from the chain scissioning of the tie chains between the crystalline and amorphous regions. As a result, the polymer fragments into segments of low molecular weight and radial strength is reduced. The hydrolytic degradation process increases pH in these regions, increasing the chain scissioning rate (autocatalysis) [48, 49]. Eventually, structural morphology of the device will show start to show cracks or discontinuities [45]. Under bulk degradation, these polymers degrade on the inside of the material and not from the surface. The hydrolyzed polymer chains are shortened and extremely hydrophilic that begins diffusing out of the device and is resorbed into the body. These by products are engulfed by phagocytes and the soluble monomers, l-lactate, are converted to pyruvate. Pyruvate enters the Krebs cycle and is then converted to carbon dioxide and water. These products are eliminated from the body by the kidneys or lungs [50].

Copolymer Degradation

Modulating the content of lactic acid to glycolic acid in a poly(lactide-glycolic acid) copolymer has shown a decrease in degradation, while the more amorphous polymers degrade faster [51, 52]. This is due to water not being able to penetrate the crystalline regions of the polymer. One group has investigated these degradation effects of porous PLGA coatings for stent [53]. They found the number and weight average molecular weight of PLGA decrease exponentially with time as shown in Figure 12.

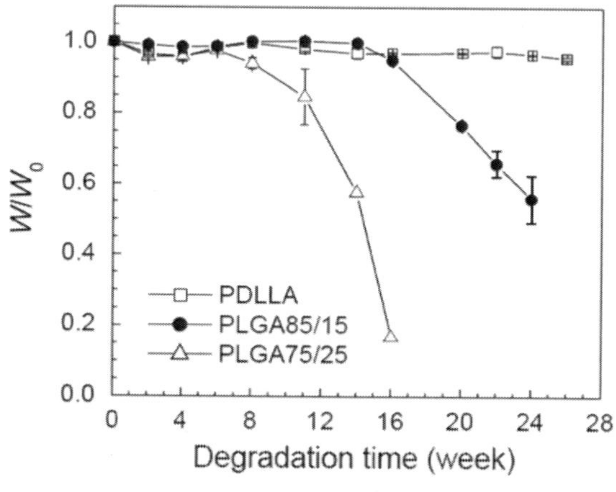

Figure 12. Shows the change in normalized weight of the scaffolds with different copolymer ratios degraded in PBS solution at 37°C. This article was published in Biomaterials, Wu and Ding, Vol.25, In vitro degradation of three-dimensional porous poly(D,L-lactide-co-glycolide) scaffolds for tissue engineering, pp. 5821-30. Copyright Elsevier (2004), with permission from Elsevier.

This can be an alternative to modulating the drug release from a coated stent. As the amount of glycolic acid groups are added to the copolymerized ratio, the degradation is accelerated with the addition of glycolic acid to the polymer chain attributed to the hydrophilicity of the lower copolymerized ratios.

Bioresorbable Stents

The development of new stent technology has been changing from metal based scaffolds to polymeric bioresorbable or biodegradable scaffolds [2]. The development of the drug eluting stents and use in the early 2000s has been associated with risk of late stent thrombosis. Redesign of these stents has included thinner struts and biodegradable coatings. These stents are permanent implants that leave an imprint within the vessel wall. The transition from metal to bioresorable technology offers several advantages being more flexible and conformability to deliver to the target lesion. They are not permanent and are resorbed within the body overtime that can help in restoration of physiological vasomotion, late luminal gain, late expansive remodeling, and reduce the risk of late stent thrombosis [54]. The stents discussed below are about some of the current polymeric designs under investigation and are shown in Figure 13.

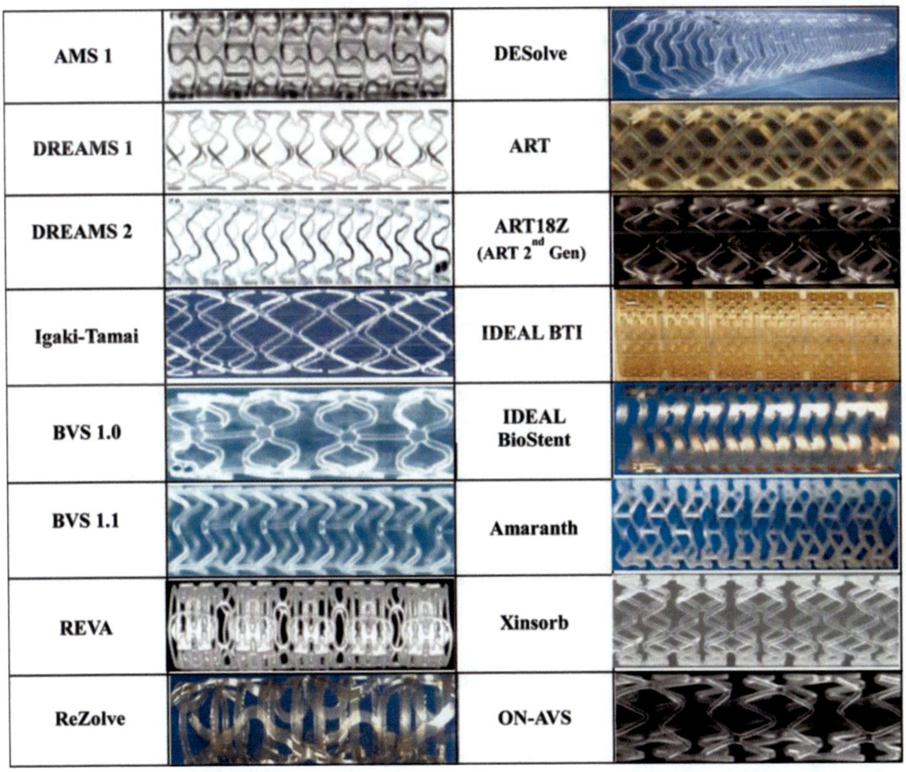

Figure 13. Bioresorbable scaffolds for preclinical and clinical use. Reprinted from Iqbal J, Onuma Y, Ormiston J, Abizaid A, Waksaman R, Serruys P, vol. 35, Bioresorbable scaffolds: rationale, current status, challenges, and future, pp. 128-31. Copyright 2013, with permission from Oxford University Press.

Duke Stent

The Duke Stent was a self-expanding stent composed of PLLA of braided filaments with a diameter of 190 μm and a Young's Modulus of 91.7×10^4 psi and a 66% elongation at break [55]. These stents were tested in the peripheral arteries of preclinical test subjects placing the stent in the femoral arteries for time points of 2 hours and 12 weeks. They reported no inflammatory response and no thrombosis and they were endothelialized after 2 weeks. This stent was not further investigated.

Iagaki-Tamai Stent

Another PLLA stent design was the Igaki-Tamai Stent that is manufactured from a monofilament fiber of varying thickness usually 170 μm with a molecular weight of 183,000 (Da) that is wound into a helical pattern [56, 57]. The stent pattern is a "zig-zag" coil that is self-expanding being deployed using a heated balloon at 70°C for 30 seconds. This higher heating temperature causes the stent to expand faster. This heating balloon is a limitation to the stent design because exposure to elevated temperatures can cause necrosis of the arterial wall and then smooth muscle cell proliferation. This temperature has been shown to promote platelet adhesion to the vessel wall and increased thrombosis. These stents are also not radiopaque and require the use of gold markers on the stent. They reported the first human 6-months result using poly(L-lactic acid) stents and the first biodegradable stent results deployed in human coronary arteries at 1 year with 10.5% restenosis and 6.7% target lesion revascularization in a total of 30 lesions (1).

BVS Stent (Abbott Labs, CA, USA)

The other PLLA stent design is the Abbott bioresorbable stent, BVS. This stent has a coating layer, PDLA, containing the antiprolerative agent everlimus. This stent is a closed cell design that has radiopaque markers on both ends of the stent [57]. The stent is a 150 μm thick strut with good mechanical properties, deliverability, and biocompatibility.

Preclinical studies on this stent have shown it is safe and effective with complete luminal endothelialization within 1 month. The ABSORB trial showed successful deployment in 94% of the patients with no late stent thrombosis, and no target lesion revascularization [58]. The stent has been approved in Europe and Asia The inflammatory response has minimal neointimal formation at 6 months with 16% luminal reduction and some incomplete apposition to the artery. The stent was fully absorbed at 2 years being integrated with fibrotic tissue [57, 59-61].

IDEAL Bioresorbable Stent (Xenogenics Corporation, MA, USA)

This stent is a tubular shaped open cell stent design composed of polylactide anhydride blended with salicylic acid and sebacic acid as a linker. The stent has a strut thickness of 200 μm and is not radiopaque. The salicylic acid is an anti-inflammatory agent however the stent is coated with salicylic acid-adipic acid with silorlimus [57, 62, 63]. The first generation required use of an 8Fr guide catheter and did not suppress proliferation due to low dosage of sirolimus. The second generation IDEAL Biostent has a higher dose with a slower release and a smaller crossing profile to deliver through a 6Fr delivery system.

ART Bioresorbable Stent (Arterial Resorbable Therapies, Paris, France)

The ART BRS is composed of PDLLA that has amorphous and semicrystalline regions in the polymers backbone chain. The design is a closed cell design with a strut thickness of 170 μm and is not radiopaque [57, 64]. The crossing profile uses a 6Fr delivery system. Preclinical studies have shown positive remodeling between 3 and 6 months. There is no coating of anti-inflammatory agents.

Amaranth FortitudeTM (Amaranth Medical, CA, USA)

The Amaranth bioresorbable stent is a cylindrical open cell design composed of PLLA with no coating. It has a strut thickness of 15-200 μm [57]. It is not radiopaque. The stent has shown good performance in bench and animal studies. They are currently in clinical studies. They have developed a sirolimus drug eluting stent that is currently under preclinical testing.

REVA (REVA Medical, CA, USA)

The REVA stent, ReZolve, is composed of desaminotyrosine polycarbonate that is bioresorbable and radio-opaque. The stent is balloon expandable with a strut thickness is 115-230 μm. Their stent consists of a "slide and lock" mechanism of teeth that are sliding that passes through a bracket. Once these teeth pass into these brackets they become locked into position. Their first generation showed no vessel recoil but high in-stent late lumen loss and target lesion revascularization. They later coated their stent with sirolimus and have being evaluated in the RESTORE trial showing good safety and efficacy at 6 months [31].

Double Opposed Helical Stent

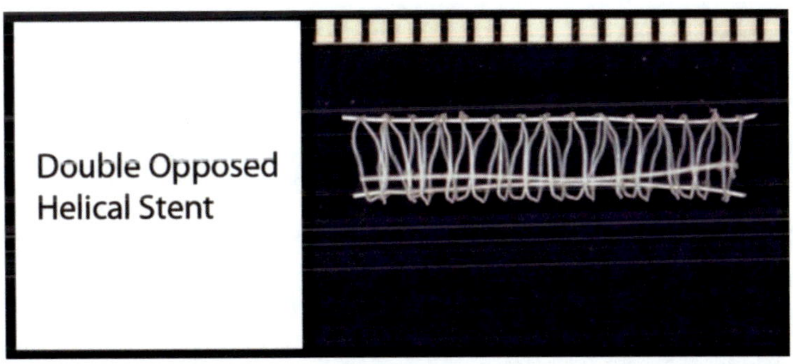

Figure 14. Showing a Double Opposed Helical Stent composed of PLLA fibers.

Using PLLA and post processing conditions a balloon expandable bioresorbable stent has been developed in Figure 14 and bench tested by Welch et al., [65] and animal tested by Veeram Reddy et al., [66]. This design consists of an intricate pattern of internal coils and axial reinforcement fibers. The internal coils have been fabricated with opposing directions to facilitate the "locking mechanism" of the stent to set the shape into an expanded cylinder. The designs range from 3mm-6mm in size and the length can be tailored to the lesion requirements usually 10-18mm length. The fiber thickness of 100-120 μm was investigated in animal studies that showed mild inflammation in rabbit models and pig models [66]. This group has successfully fabricated larger stent sizes up to 12mm diameter using thicker fibers of 200-250 μm.

Self-Expanding PLLA/PLGA Bilayer Stent

A combination of PLLA and PLGA layers for stent fabrication to have a truly self-expanding stent has been investigated. Venkatraman et al., [67] were able to use the concept of partial elastic memory to create a bilayer of PLGA on a base material of PLLA. These materials are heat treated just above their glass transition temperatures and annealed on a mandrel of their specific target stent diameter. These stents are a spiral-coil shaped and composed of thickness ranges from 50-150 μm. They are then crimped and permitted to self-expand at 37°C. Their work showed that using a bilayer approach they were able to create a self-expanding stent but more investigational work was needed to address collapse strength, inflammatory response and degradation kinetics.

Conclusion

PLLA and PLGA bioresorbable materials have been investigated not only as stent design but as coating materials for controlled drug release. In this chapter, we described bioresorbable stent designs and stent platforms that use these bioresorbable coatings for drug eluting stents. The focus was targeted to PLGA and PLLA polymers. Other inorganic coatings have been investigated and metal degrading stents, Biotroniks AMS-1, DREAMS-1 and Dreams-2 designs.

References

[1] Tamai H, Igaki K, Kyo E, Kosuga K, Kawashima A, Matsui S, et al. Initial and 6-month results of biodegradable poly-l-lactic acid coronary stents in humans. *Circulation*. 2000;102(4):399-404.

[2] Garg S, Serruys PW. Coronary stents: looking forward. *Journal of the American College of Cardiology*. 2010;56(10 Suppl):S43-78.

[3] Ormiston JA, Serruys PW, Regar E, Dudek D, Thuesen L, Webster MW, et al. A bioabsorbable everolimus-eluting coronary stent system for patients with single de-novo coronary artery lesions (ABSORB): a prospective open-label trial. *Lancet*. 2008;371(9616):899-907.

[4] Diletti R, Serruys PW, Farooq V, Sudhir K, Dorange C, Miquel-Hebert K, et al. ABSORB II randomized controlled trial: a clinical evaluation to compare the safety, efficacy, and performance of the Absorb everolimus-eluting bioresorbable vascular scaffold system against the XIENCE everolimus-eluting coronary stent system in the treatment of subjects with ischemic heart disease caused by de novo native coronary artery lesions: rationale and study design. *American heart journal*. 2012;164(5):654-63.

[5] Wang X, Venkatraman SS, Boey FY, Loo JS, Tan LP. Controlled release of sirolimus from a multilayered PLGA stent matrix. *Biomaterials*. 2006;27(32):5588-95.

[6] Zilberman M. Dexamethasone loaded bioresorbable films used in medical support devices: structure, degradation, crystallinity and drug release. *Acta biomaterialia*. 2005;1(6):615-24.

[7] Finkelstein A, McClean D, Kar S, Takizawa K, Varghese K, Baek N, et al. Local drug delivery via a coronary stent with programmable release pharmacokinetics. *Circulation.* 2003;107(5):777-84.

[8] Tan A, Farhatnia Y, de Mel A, Rajadas J, Alavijeh MS, Seifalian AM. Inception to actualization: next generation coronary stent coatings incorporating nanotechnology. *Journal of biotechnology*. 2013;164(1):151-70.

[9] Park KW, Chae IH, Lim DS, Han KR, Yang HM, Lee HY, et al. Everolimus-eluting versus sirolimus-eluting stents in patients undergoing percutaneous coronary intervention: the EXCELLENT (Efficacy of Xience/Promus Versus Cypher to Reduce Late Loss After Stenting) randomized trial. *Journal of the American College of Cardiology*. 2011;58(18):1844-54.

[10] Xi T, Gao R, Xu B, Chen L, Luo T, Liu J, et al. In vitro and in vivo changes to PLGA/sirolimus coating on drug eluting stents. *Biomaterials*. 2010;31(19):5151-8.

[11] Morice MC, Serruys PW, Sousa JE, Fajadet J, Ban Hayashi E, Perin M, et al. A randomized comparison of a sirolimus-eluting stent with a standard stent for coronary revascularization. *The New England journal of medicine*. 2002;346(23):1773-80.

[12] Leon MB, Mauri L, Popma JJ, Cutlip DE, Nikolsky E, O'Shaughnessy C, et al. A randomized comparison of the Endeavor zotarolimus-eluting stent versus the TAXUS paclitaxel-eluting stent in de novo native coronary lesions 12-month outcomes from the ENDEAVOR IV trial. *Journal of the American College of Cardiology*. 2010;55(6):543-54.

[13] Lotan C, Meredith IT, Mauri L, Liu M, Rothman MT, Investigators EF. Safety and effectiveness of the Endeavor zotarolimus-eluting stent in real-world clinical practice: 12-month data from the E-Five registry. *JACC Cardiovascular interventions*. 2009;2(12):1227-35.

[14] Eisenstein EL, Wijns W, Fajadet J, Mauri L, Edwards R, Cowper PA, et al. Long-Term Clinical and Economic Analysis of the Endeavor Drug-Eluting Stent Versus the Driver Bare-Metal Stent4-Year Results From the ENDEAVOR II Trial (Randomized Controlled Trial to Evaluate the Safety and Efficacy of the Medtronic AVE ABT-578 Eluting Driver Coronary Stent in De Novo Native Coronary Artery Lesions*). JACC: Cardiovascular Interventions*. 2009;2(12):1178-87.

[15] Lee BK, Yun YH, Choi JS, Choi YC, Kim JD, Cho YW. Fabrication of drug-loaded polymer microparticles with arbitrary geometries using a piezoelectric inkjet printing system. *International journal of pharmaceutics*. 2012;427(2):305-10.

[16] Verkouteren RM, Verkouteren JR. Inkjet metrology II: resolved effects of ejection frequency, fluidic pressure, and droplet number on reproducible drop-on-demand dispensing. Langmuir : *the ACS journal of surfaces and colloids*. 2011;27(15):9644-53.

[17] Tarcha P, Verlee D, Hui H, Setesak J, Antohe B, Radulescu D, et al. The Application of Ink-Jet Technology for the Coating and Loading of Drug-Eluting Stents. *Annals of biomedical engineering*. 2007;35(10):1791-9.

[18] Jang D, Kim D, Moon J. Influence of fluid physical properties on ink-jet printability. Langmuir : *the ACS journal of surfaces and colloids*. 2009;25(5):2629-35.

[19] Perkins J, Hong Y, Ye SH, Wagner WR, Desai S. Direct writing of bio-functional coatings for cardiovascular applications. *Journal of biomedical materials research Part* A. 2014.

[20] Chen MC, Liang HF, Chiu YL, Chang Y, Wei HJ, Sung HW. A novel drug-eluting stent spray-coated with multi-layers of collagen and sirolimus. *Journal of controlled release : official journal of the Controlled Release Society.* 2005;108(1):178-89.

[21] Bege N, Steinmuller SO, Kalinowski M, Reul R, Klaus S, Petersen H, et al. Drug eluting stents based on Poly(ethylene carbonate): optimization of the stent coating process. *European journal of pharmaceutics and biopharmaceutics : official journal of Arbeitsgemeinschaft fur Pharmazeutische Verfahrenstechnik eV.* 2012;80(3):562-70.

[22] Jabara R, Chronos N, Conway D, Molema W, Robinson K. Evaluation of a novel slow-release paclitaxel-eluting stent with a bioabsorbable polymeric surface coating. *JACC Cardiovascular interventions.* 2008;1(1):81-7.

[23] Stone GW, Ellis SG, Colombo A, Grube E, Popma JJ, Uchida T, et al. Long-term safety and efficacy of paclitaxel-eluting stents final 5-year analysis from the TAXUS Clinical Trial Program. *JACC Cardiovascular interventions.* 2011;4(5):530-42.

[24] Ranade SV, Miller KM, Richard RE, Chan AK, Allen MJ, Helmus MN. Physical characterization of controlled release of paclitaxel from the TAXUS Express2 drug-eluting stent. *Journal of biomedical materials research Part A.* 2004;71(4):625-34.

[25] Pinchuk L, Wilson GJ, Barry JJ, Schoephoerster RT, Parel JM, Kennedy JP. Medical applications of poly(styrene-block-isobutylene-block-styrene) ("SIBS"). *Biomaterials.* 2008;29(4):448-60.

[26] Venkatraman S, Boey F. Release profiles in drug-eluting stents: issues and uncertainties. *Journal of controlled release : official journal of the Controlled Release Society.* 2007;120(3):149-60.

[27] Chieffo A, Colombo A. Polymer-based paclitaxel-eluting coronary stents. Clinical results in de novo lesions. *Herz.* 2004;29(2):147-51.

[28] Grube E, Silber S, Hauptmann KE, Mueller R, Buellesfeld L, Gerckens U, et al. TAXUS I: six- and twelve-month results from a randomized, double-blind trial on a slow-release paclitaxel-eluting stent for de novo coronary lesions. *Circulation.* 2003;107(1):38-42.

[29] Stone GW, Ellis SG, Cox DA, Hermiller J, O'Shaughnessy C, Mann JT, et al. One-year clinical results with the slow-release, polymer-based, paclitaxel-eluting TAXUS stent: the TAXUS-IV trial. *Circulation.* 2004;109(16):1942-7.

[30] Grube E, Buellesfeld L. BioMatrix Biolimus A9-eluting coronary stent: a next-generation drug-eluting stent for coronary artery disease. *Expert review of medical devices.* 2006;3(6):731-41.

[31] Costa RA, Lansky AJ, Abizaid A, Mueller R, Tsuchiya Y, Mori K, et al. Angiographic results of the first human experience with the Biolimus A9 drug-eluting stent for de novo coronary lesions. *The American journal of cardiology.* 2006;98(4):443-6.

[32] Grube E, Hauptmann KE, Buellesfeld L, Lim V, Abizaid A. Six-month results of a randomized study to evaluate safety and efficacy of a Biolimus A9 eluting stent with a biodegradable polymer coating. EuroIntervention : journal of EuroPCR in collaboration with the Working Group on Interventional Cardiology of the European Society of Cardiology. 2005;1(1):53-7.

[33] Berger HL. Using ultrasonic spray nozzles to coat drug-eluting stents. *Medical device technology.* 2006;17(9):44-6, 8-9.

[34] Lang RJ. Ultrasonic atomization of liquids. *Journal of Acoustic Society America.* 1962;34:6-8.

[35] Wood WR, Loomis AL. The physical and biological effects of high frequency sound-waves of great intensity. *Phil Mag*. 1927;4:417-37.

[36] Pan CJ, Tang JJ, Weng YJ, Wang J, Huang N. Preparation, characterization and anticoagulation of curcumin-eluting controlled biodegradable coating stents. *Journal of Controlled Release*. 2006;116(1):42-9.

[37] Shanshan C, Lili T, Yingxue T, Bingchun Z, Ke Y. Study of drug-eluting coating on metal coronary stent. Materials science & engineering C, *Materials for biological applications*. 2013;33(3):1476-80.

[38] Pan CJ, Tang JJ, Weng YJ, Wang J, Huang N. Preparation and characterization of rapamycin-loaded PLGA coating stent. *Journal of materials science Materials in medicine*. 2007;18(11):2193-8.

[39] Pan CJ, Tang JJ, Weng YJ, Wang J, Huang N. Preparation and in vitro release profiles of drug-eluting controlled biodegradable polymer coating stents. *Colloids and surfaces B, Biointerfaces*. 2009;73(2):199-206.

[40] Pan CJ, Wang J, Huang N. Preparation, characterization and in vitro anticoagulation of emodin-eluting controlled biodegradable stent coatings. *Colloids and surfaces B, Biointerfaces*. 2010;77(2):155-60.

[41] Carlyle WC, McClain JB, Tzafriri AR, Bailey L, Zani BG, Markham PM, et al. Enhanced drug delivery capabilities from stents coated with absorbable polymer and crystalline drug. *Journal of controlled release : official journal of the Controlled Release Society*. 2012;162(3):561-7.

[42] Tomlins P. Influence of porous structure on bioresorbablitiy: Tissue engineering scaffolds. In: Buchanan F, editor. Degradation rate of bioresorbable materials Prediction and Evaluation. Cambridge, England: Woodhead Publishingand Maney Publishihing; 2008. p. 234-57.

[43] Pitt CG, Gratzl MM, Kimmel GL, Surles J, Schindler A. Aliphatic polyesters II. The degradation of poly (DL-lactide), poly (epsilon-caprolactone), and their copolymers in vivo. *Biomaterials*. 1981;2(4):215-20.

[44] Pitt CG, Zhong-wei G. Modification of the rates of chain cleavage of poly(ϵ-caprolactone) and related polyesters in the solid state. *Journal of Controlled Release*. 1987;4(4):283-92.

[45] Weir NA, Buchanan FJ, Orr JF, Dickson GR. Degradation of poly-L-lactide. Part 1: in vitro and in vivo physiological temperature degradation. Proceedings of the Institution of Mechanical Engineers Part H, *Journal of engineering in medicine*. 2004;218(5):307-19.

[46] Weir NA, Buchanan FJ, Orr JF, Farrar DF, Dickson GR. Degradation of poly-L-lactide. Part 2: increased temperature accelerated degradation. Proceedings of the Institution of Mechanical Engineers Part H, *Journal of engineering in medicine*. 2004;218(5):321-30.

[47] Oberhauser JP, Hossainy S, Rapoza RJ. Design principles and performance of bioresorbable polymeric vascular scaffolds. EuroIntervention : *journal of EuroPCR in collaboration with the Working Group on Interventional Cardiology of the European Society of Cardiology*. 2009;5 Suppl F:F15-22.

[48] Hyon SH, Jamshidi K, Ikada Y. Effects of residual monomer on the degradation of DL-lactide polymer. *Polymer International*. 1998;46(3):196-202.

[49] Li S, McCarthy S. Further investigations on the hydrolytic degradation of poly (DL-lactide). *Biomaterials*. 1999;20(1):35-44.

[50] Hollinger JO, Battistone GC. Biodegradable bone repair materials. Synthetic polymers and ceramics. *Clinical orthopaedics and related research*. 1986(207):290-305.

[51] Miller RA, Brady JM, Cutright DE. Degradation rates of oral resorbable implants (polylactates and polyglycolates): rate modification with changes in PLA/PGA copolymer ratios. *Journal of biomedical materials research*. 1977;11(5):711-9.

[52] Wu XS, Wang N. Synthesis, characterization, biodegradation, and drug delivery application of biodegradable lactic/glycolic acid polymers. Part II: biodegradation. *Journal of biomaterials science Polymer edition*. 2001;12(1):21-34.

[53] Wu L, Ding J. In vitro degradation of three-dimensional porous poly(D,L-lactide co glycolide) scaffolds for tissue engineering. *Biomaterials*. 2004;25(27):5821-30.

[54] Serruys PW, Garcia-Garcia HM, Onuma Y. From metallic cages to transient bioresorbable scaffolds: change in paradigm of coronary revascularization in the upcoming decade? *European heart journal*. 2012;33(1):16-25b.

[55] Stack RS, Califf RM, Phillips HR, Pryor DB, Quigley PJ, Bauman RP, et al. Interventional cardiac catheterization at Duke Medical Center. *The American journal of cardiology*. 1988;62(10 Pt 2):3f-24f.

[56] Tamai H, Igaki K, Tsuji T, Kyo E, Kosuga K, Kawashima A, et al. A Biodegradable Poly-l-lactic Acid Coronary Stent in the Porcine Coronary Artery. *Journal of Interventional Cardiology*. 1999;12(6):443-50.

[57] Iqbal J, Onuma Y, Ormiston J, Abizaid A, Waksman R, Serruys P. Bioresorbable scaffolds: rationale, current status, challenges, and future. *European heart journal*. 2014;35(12):765-76.

[58] Ormiston JA, Webster MW, Armstrong G. First-in-human implantation of a fully bioabsorbable drug-eluting stent: the BVS poly-L-lactic acid everolimus-eluting coronary stent. *Catheterization and cardiovascular interventions : official journal of the Society for Cardiac Angiography & Interventions*. 2007;69(1):128-31.

[59] Moore J, Jr., Soares J, Rajagopal K. Biodegradable Stents: Biomechanical Modeling Challenges and Opportunities. *Cardiovasc Eng Tech*. 2010;1(1):52-65.

[60] Serruys PW, Ormiston JA, Onuma Y, Regar E, Gonzalo N, Garcia-Garcia HM, et al. A bioabsorbable everolimus-eluting coronary stent system (ABSORB): 2-year outcomes and results from multiple imaging methods. *Lancet*. 2009;373(9667):897-910.

[61] Garcia-Garcia HM, Gonzalo N, Pawar R, Kukreja N, Dudek D, Thuesen L, et al. Assessment of the absorption process following bioabsorbable everolimus-eluting stent implantation: temporal changes in strain values and tissue composition using intravascular ultrasound radiofrequency data analysis. A substudy of the ABSORB clinical trial. EuroIntervention : journal of EuroPCR in collaboration with the Working Group on Interventional Cardiology of the European Society of Cardiology. 2009;4(4):443-8.

[62] Jabara R, Chronos N, Robinson K. Novel bioabsorbable salicylate-based polymer as a drug-eluting stent coating. *Catheterization and cardiovascular interventions : official journal of the Society for Cardiac Angiography & Interventions*. 2008;72(2):186-94.

[63] Jabara R, Pendyala L, Geva S, Chen J, Chronos N, Robinson K. Novel fully bioabsorbable salicylate-based sirolimus-eluting stent. *EuroIntervention : journal of EuroPCR in collaboration with the Working Group on Interventional Cardiology of the European Society of Cardiology*. 2009;5 Suppl F:F58-64.

[64] Lafont A, Durand E. A.R.T.: concept of a bioresorbable stent without drug elution. EuroIntervention : *journal of EuroPCR in collaboration with the Working Group on Interventional Cardiology of the European Society of Cardiology*. 2009;5 Suppl F: F83-7.

[65] Welch T, Eberhart R, Reddy S, Wang J, Nugent A, Forbess J. Novel Bioresorbable Stent Design and Fabrication: Congenital Heart Disease Applications. *Cardiovasc Eng Tech*. 2013:1-12.

[66] Veeram Reddy SR, Welch TR, Wang J, Bernstein F, Richardson JA, Forbess JM, et al. A novel biodegradable stent applicable for use in congenital heart disease: bench testing and feasibility results in a rabbit model. *Catheterization and cardiovascular interventions : official journal of the Society for Cardiac Angiography & Interventions*. 2014;83(3):448-56.

[67] Venkatraman SS, Tan LP, Joso JF, Boey YC, Wang X. Biodegradable stents with elastic memory. *Biomaterials*. 2006;27(8):1573-8.

In: New Developments in Polylactic Acid Research
Editor: Courtney Winthrop
ISBN: 978-1-63463-054-2

Chapter 2

Controlled Drug Delivery Systems Based on PLA

Vicente Rives, Patricia H. Holgado, María Jesús Holgado and María Soledad San Román
GIR-QUESCAT, Departamento de Química Inorgánica, Universidad de Salamanca, Salamanca, Spain

Abstract

Biodegradable polymers are used in Medicine to replace and increase the volume of tissues, as tissue supports and also for controlled drug release. The most widely used biodegradable polymers are polylactic acid (PLA), polyglycolic acid (PGA), and ε-caprolactone and their copolymers, as they are easily biodegraded. These materials are used as microspheres or microcapsules to host a broad diversity of drugs, such as antiinflamatory, anticonceptives, narcotics antagonists, local anesthetics and vaccines. Release of drugs from these systems follows different mechanisms, including diffusion and hydrolytic or enzimatic degradation of the polymer. Biodegradability depends on the chemical structure of the material and on the composition of the final product. In many cases a fast degradation is required and no polymer is used; a slower release is observed when using PLA, and the combined use of PLA and PGA copolymers permits to modulate the release degree. The ability of the polymer to control release of active molecules depends on its degradation profile: on controlling the degradation rate of the polymer degradation we can control the release of the hosted active molecule.

Other family of compounds which are currently also being used for controlled release of drugs are layered double hydroxides, LDH, with the hydrotalcite-like structure. These compounds are able to encapsulate the drugs. Their structure consists of stacked layers of $[M^{II}(OH)_6]$ octahedra sharing edges, where a partial M^{II}/M^{III} isomorphic substitution has taken place, and the positive charge in excess is balanced by anions intercalated between the sheets. These anions can be exchanged, introducing anionic drugs, or the solid can be even directly synthetized with the drug within the interlayer space. Formation of LDH/drug systems increases the solubility of the drug, affecting its effect on the living organism and the ability for being transported in the body fluids. The solubility is also related to release, absorption and bioassimilation of the drug.

The nanocompounds which study is here reported can be considered organic-inorganic nanohybrids integrated in polymers forming nanocomposites. LDHs are used to host a given drug and then this organic-inorganic nanohybrid system is supported on a biodegradable polymer, such as PLA, which sometimes induces a partial exfoliation of the LDH sheets. PLA markedly delays release of the drug from several hours when using a LDH-drug system, to several weeks or even months in the presence of PLA. We have summarized the current knowledge on these systems, paying special attention to the effect of different preparation procedures on the physicochemical properties (characterization) and release ability.

List of Abbreviations

5ASA	5-aminosalicylic acid
5FU	5-fluorouracil
AMOX	Amoxycillin
ASO	Antisense oligonucleotide
Bz	Benzoate
BzDc	2,4-Dichlorobenzoate
C12	Laurate
C18	Estearate
CDD	Controlled Drug Delivery
CHLO	Chloramphenicol
DBS	Dodecyl benzene sulfonate
DICLO	Diclofenac
DOC	Docetaxel
DOX	Doxorubicin
DS	Dodecyl sulphate
DTA	Differential thermal analysis
DTG	Derivative thermogravimetry
EUD	Eudragit
EVA	Ethylene vynil acetate
EXAFS	Extended X-ray absorption fine structure
FENB	Fenbufen
FTIR	Fourier transform infrared spectroscopy
GA	Glycolic acid
GTR	Guided tissue regeneration
HD	12-hydroxydodecanoate
HEBM	High-energy ball milling
HPLC	High performance liquid chromatography
IAA	Indole-3-acetic acid
IBU	Ibuprofen
IND	Indomethacin
KET	Ketoprofen
LA	Lactic acid
LBL	Layer by layer
LDH	Layered double hydroxide

MW	Microwave
NAA	Naphthaleneacetate
NAP	Naproxene
NC	Nanocomposite
NMR	Nuclear magnetic resonance
NSAID	Non steroidal antiinflammatory drug
o-BzOH	orto-hydroxybenzoate
PBS	Phosphate buffered saline
p-BzOH	para-hydroxybenzoate
PA6	Polyamide-6
PCL	Poly(ε-caprolactone)
PEG	Poly (ethylene glycol)
PET	Poly (ethylene terephthalate)
PGA	Poly (glycolic acid)
PHA	Polyhydroxyalcanoate
PLA	Poly (lactic acid)
PLGA	Poly(DL-lactic acid-co-glycolic acid) copolymer
PLURONIC	Triblock copolymer based on poly(ethylene glycol)-poly(propylene glycol)-poly(ethylene glycol)
PXRD	Powder X-ray diffraction
PVC	Poly vinyl chloride
ROP	Ring opening polymerization
SEM	Scanning electron microscopy
SF	Sodium fusidate
SGF	Simulated gastric fluid
SIF	Simulated intestinal fluid
TEC	Triethyl citrate
TEM	Transmission electron microscopy
TET	Tetracycline
TG	Thermogravimetric analysis
THF	Tetrahydrofurane
UDC	Ursodesoxicolate
UV-Vis	Ultraviolet visible spectroscopy
VAN	Vancomycin
XANES	X-ray absorption near edge structure
XAS	X-ray absorption spectroscopy

Introduction

The use of organic polymers has increased drastically in the last decades. However, current concern about environmental protection has led to the use of *biodegradable* polymers, which do not accumulate in the environment, and can be considered as typical products of the so-called *Green Chemistry* [1-3]. New and interesting applications are periodically reported for these biodegradable polymers, but specially in Biomedicine. A new set of materials for therapeutical use has been developed, made with these polymers, with different biomedical functions, such as mechanical support, mechanical barrier, man-made tissues and organs, cell supports, as well as pharmacological uses as drug supports and drug carriers and systems for controlled drug delivery (CDD) [4,5]. Despite these functions can be also performed by non-biodegradable compounds, the biodegradable ones are preferred, as they will remain in the organism unless they are quirurgically removed. Due to the problems for small scale recovery of these components, the us of biodegradable polymers is even more advantageous, as they are naturally degradated (biodegradated) and disappear from the tissue after a previously determined period of time [6,7].

Man-made polymers usually show better performance than the natural ones, as they are tailored-synthetized for the specific aim they are designed to and will be used; for instance, different polymers are more suitable for tissue engineering than for drug delivery, and also they are less prone to produce toxic effects. To adjust the degradation rate to the desired target, copolymers are sometimes used, well mixed or copolymerized [8]. Consequently, high versatile polymers can (should) be prepared. On the other side, on controlling the degradation process we can control the release of molecules trapped within the polymer network, and to fit this release (rate and concentration) to the specific therapeutical needs.

In addition to polymers, other materials have been used as drug carriers for CDD. Usually high specific surface area, inert, inorganic compounds, as clays or layered double hydroxides (LDHs). So, we have the enormous amount of knowledge on drug/polymer systems and on drug/inorganic carrier systems and several research teams have proposed a hybrid approach, by using drugs supported on inorganic carriers, dispersed in a polymer matrix, aiming to reach a synergic scenario where the advantages of both systems meet and improve to each other.

We here report on the use of drugs hosted in a layered double hydroxide matrix, and dispersed in polylactic acid. Our approach has been the following: First a short description of polylactic acid (PLA), one of the biodegradable polymers most widely used in biomedical and pharmacological applications, paying special attention to its degradation, as this is the mechanism by which the drug is released. Secondly we describe the structure, synthesis and applications of layered double hydroxides, a family of compounds which have shown their good performance in different applications in many fields, Medicine and Pharmacology being one of them. Then we describe the use of PLA as a drug host and carrier, and discuss the mechanisms through which the drug is released, describing also the different approaches reported for controlling this release. The next section describes the same sort of applications, but using LDHs, and not PLA, as carriers; this is a rather short section, as the subject has been recently reviewed by some of us [9,10]. The use of the three-components nanocomposites, drug-LDH-PLA, is discussed in the next section, and the Chapter ends with a description of the works reported using other polymers and copolymers to support drug-LDH systems.

Polylactic Acid and Copolymers

We call biodegradable materials to those able to be degraded environmentally. They can be taken as a new generation of materials which use will significantly reduce their impact on the environment in terms of energy consumption and generation of residues when used.

Biodegradability of plastics depends on their chemical composition and structure, not only on the nature of the reactants used to prepare them [11]; for this reason, biodegradable materials can be prepared from natural or synthetic resins. Natural biodegradable plastics are usually prepared from renewable resources, such as scarch, while synthetic resins are prepared from oil derivatives and other products; examples are polyesters and polyethylene. As all commercially and industrially interesting products, biodegradable materials should fulfill some requirements, which depend on the use they are intended for. For this reason, biodegradable plastics are often mixed with synthetic polymers to improve their functional properties, as demanded by markets.

Biodegradable polymers can be classified in four categories, mainly related to their origin and production:

- Polymers which proceed from natural sources; they are obtained from sea beings or easily accessible vegetals; scarch, cellulose, casein and gluten belong to this group.
- Polymers obtained by classical chemical synthesis or from renewable biological monomers. The best example in this group is polylactic acid (PLA), a biopolyester obtained from lactic acid.
- Polymers produced by microorganisms or genetically modified bacteriae. Examples are polyhydroxyalcanoates (PHAs)
- Finally, synthetic polymers prepared from oil, such as polycaprolactone (PCL).

Polylactic Acid (PLA)

This is a synthetic, biodegradable, polymer classified as an α-hydroxyacid or aliphatic polyester, first described by Scheele in 1780 [12,13]. It was synthetized by Pelouze in 1845 by condensation of lactic acid [14]. Avery patented a process consisting of fermentation of vegetal carbohydrates to produce PLA [15]. In 1932, Carothers et al. developed a method to polymerize lactide to produce PLA, which was patented by DuPont in 1954 [16]. Due to its high price, PLA was originally restricted to uses in biomedical applications, although the relatively low molecular mass of the polymers restricted their applications. An important advance was reported in 1990, when high molecular mass PLA was prepared from commercial lactide [17,18], which was marketed in 1997 as NatureWorks, starting large scale production and use of PLA.

The possibility to produce large amounts of PLA has also launched its use in research and development; moreover, the concerns about the environment and development of what is now known as Green Chemistry encouraged the use of biological-based polymers. The key factor in the use of renewable plastic materials is the economic sostenibility. Consequently, to succeed in production of PLA the cost to obtain it should be competitive. The industrial, low-cost, process to produce PLA involves a production process based on fermentation to produce lactic acid from an efficient use of carbohydrates; in a second step a very effective purification process avoids the presence of by-products. The PLA market is able to absorb the world wide production of lactic acid derived from sugar fermentation, using carbohydrate sources such as corn, sugarcane or tapioca. The structure of PLA is shown in Figure 1.

Synthesis of PLA

Briefly, PLA is produced by elimination of water from lactic acid by condensation. A lactide (a cyclic dimer of lactic acid formed by dehydration) ring is first formed by elimination of two water molecules by condensation of two molecules of lactic acid; polymerization takes place upon opening of the ring.

The synthesis can be also carried out by direct condensation of lactic acid after opening the ring of L-lactide (ROP, *ring opening polymerization*). This method permits an easier control of the reaction and thus the properties of the final polymer can be controlled [19,20]. Once the equilibrium is reached, it is usually difficult to remove water in the last steps of the polymerization, and thus the molecular mass does not reach very high values; this is undoubtedly a drawback of this method.

Cargill Dow LLC have developed a cheap, continous process to produce lactic acid-based polymers [21]. First a continuous condensation of lactic acid in an aqueous medium takes place, producing a low molecular mass prepolymer. This prepolymer is then converted into a mixture of stereoisomers of the lactide by using a tin-based catalyst, *via* a rather selective intramolecular cyclation. The lactide mixture is purified by distillation under vacuum. Finally, high molecular mass PLA is formed by ROP in the presence of the tin catalyst. The main advantage of this method is that it avoids the use of expensive and highly contaminating organic solvents. A scheme of the condensation-ROP method to produce PLA is depicted in Figure 2.

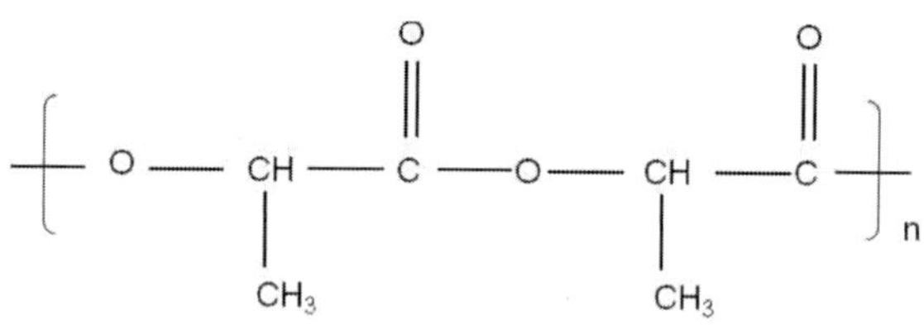

Figure 1. Chemical formula of a fragment of poly lactic acid, PLA.

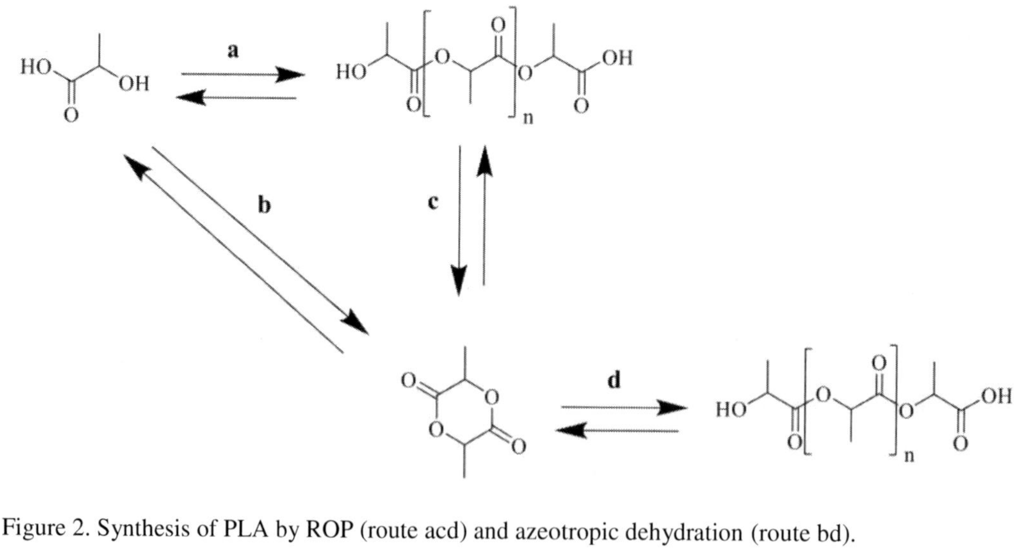

Figure 2. Synthesis of PLA by ROP (route acd) and azeotropic dehydration (route bd).

Despite the ROP method is very widely used, Mitsui Toatshu Chemical have patented [22-24] a method known as *azeotropic dehydration*, using a solvent (anisole or diphenyl ether), which permits a direct esterification process and to obtain high molecular mass PLA. The main reactions involved in this process are shown in Figure 2.

Degradation of PLA

The first medical application of PLA was reported by Kulkarni et al. [25] in 1966. In addition to being biodegradable, it is also biocompatible, avoiding rejection by the organism and formation of fibrous capsules and infections. The biocompatible and biodegradable compounds remain in the body for a limited period of time (which can be extend along months) and disappear upon biodegradation. To fit the time required for the biodegradation and the time required by the tissues surrounding the polymer to heal is one of the aims in design of biodegradable implantations [26]. Microspheres and microcapsules have been widely used in drug delivery systems of a broad extension of drugs, such as anticonceptives, narcotic antagonists, local anesthesia, vaccins, etc. Release of drugs from these systems is based on several mechanisms, including diffusion and degradation of the polymer (hydrolylitic or enzymatic degradation). Again, the time fitting is crucial: a fast degradation gives rise to a burst increase of drug which can produce toxic situations and the drug is misused by the organism; on the contrary, if the release is too slow the therapeutical drug concentration is not reached. PLA usually degrades very slowly [27], while poly(DL-lactic acid co glycolic acid) copolymer (PLGA) or mixtures of PLA with other polymers can provide a faster degradation.

The main problem found with PLA matrices is the low degradation degree even for the large amounts of residues formed. Degradation can take place *via* three different routes: hydrolytic, thermal and enzymatic.

Hydrolytic Degradation

Hydrolytic degradation depends on the molecular mass and structure of the polymer, as well as temperature and pH, and the catalysts (alkaline species, enzymes, etc.) existing in the medium. When used in biomedical and pharmaceutical applications, degradation of PLA should be controlled *via* these factors; the specific PLA-based material should be selected and conditioned for an optimum degradation rate to maintain a constant release of the drug to the medium.

Hydrolytic degradation of PLA takes place through breaking of the ester group, forming hydroxyl and carboxylic groups. The process is affected by the reaction medium. In an acidic medium lactic acid and its oligomers are formed; however, in a basic medium lactide cyclic units are first formed, although they are then opened, producing lactoyllactic acid, which is hydrolyzed again, finally leading to formation of lactic acid, Figure 3.

The hydrolytic reaction leads to formation of water soluble, low molecular mass monomers and oligomers, *via* breaking of the chains, leading to a decrease in the molecular mass of the remaining chain. The degradation degree can be determined by HPLC or NMR. The hydrolysis rate also depends on the crystallinity degree of the solid, its specific surface area and its character, originally hydrophobic, but it can be changed to increase the hydrophilic character. Crystalline PLA is more resistant than amorphous PLA to degradation, and in mixed systems degradation affects more strongly to the amorphous than to the crystalline parts.

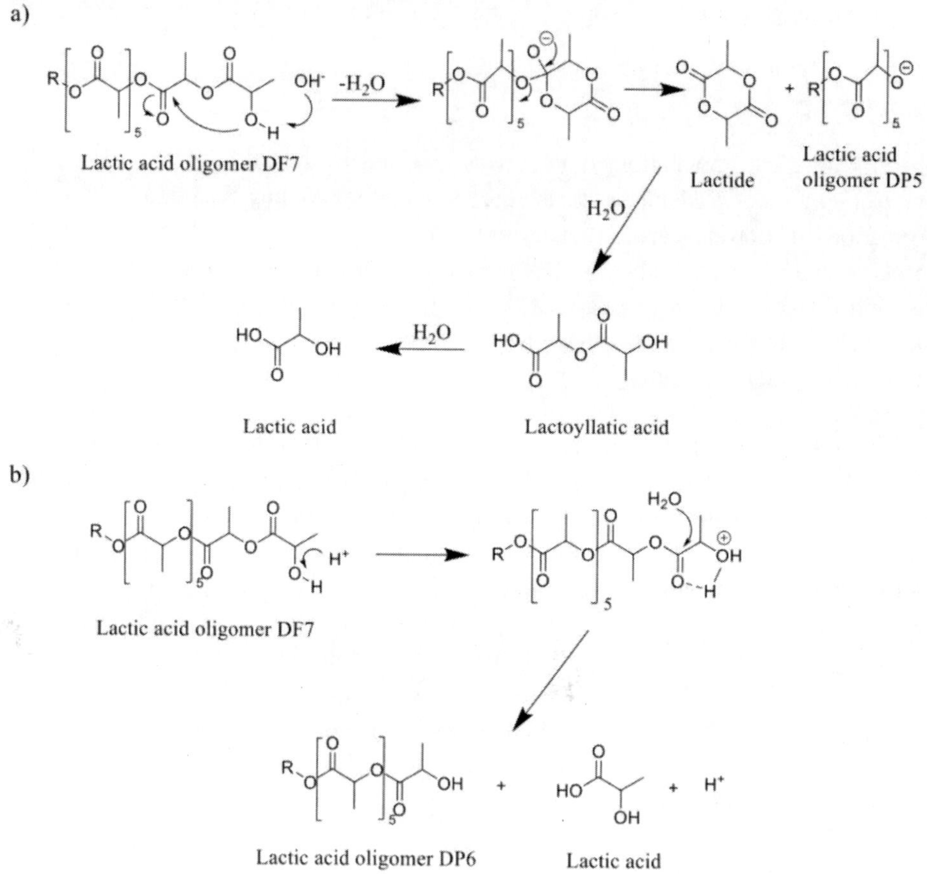

Figure 3. Degradation of PLA in(a) alkaline, and (b) acid media.

Hydrolytic degradation has also an erosion effect on the surface of the PLA particles in contact with water. This process is more severe in the presence of catalysts (alkaline agents and some enzymes); the hydrolytic degradation seems to affect more effectively to the surface of the particles, as there it is much faster than in the core of the particles [28]. The thickness of the particles is also important, and surface erosion is larger when the thickness is larger than a *critical thickness*. The PLA chains in the crystalline region are resistant to hydrolysis in comparison to those in the amorphous parts, due to the hindered access of water molecules inside the crystalline regions. As a result, selective breakdown of bonds induced by water in the amorphous region forming water soluble monomers and oligomers is favoured, resulting in a selective hydrolytic degradation, removing chains from the amorphous region, while the crystalline parts remain almost unaltered [29].

Hydrolytic degradation is also enhanced by a temperature increase, and the molecular mass of the fractions formed decreases in the same sense, Figure 4.

Enzymatic Degradation

Water insoluble biodegradable plastics can be degraded to soluble compounds (organic acids or oligomers) by enzymes secreted by microorganisms; the degradation compounds can be metabolized finally forming CO_2, H_2O and biomass. PLA can be degraded by enzymes such as pronase, proteinase K and bromelain.

The process takes place in two steps. In the first one, the long, high molecular mass PLA chain is degraded to lower molecular mass oligomers. This process can be accelerated by acids or bases, and its rate depends on the temperature and the humidity. Fragmentation of the chains takes place in this step, approximately when the molecular mass decreases to values below 40 kDa; from this point onwards the degradation process is carried out by microorganisms in the organic medium, finally forming CO_2, H_2O and biomass [30,31].

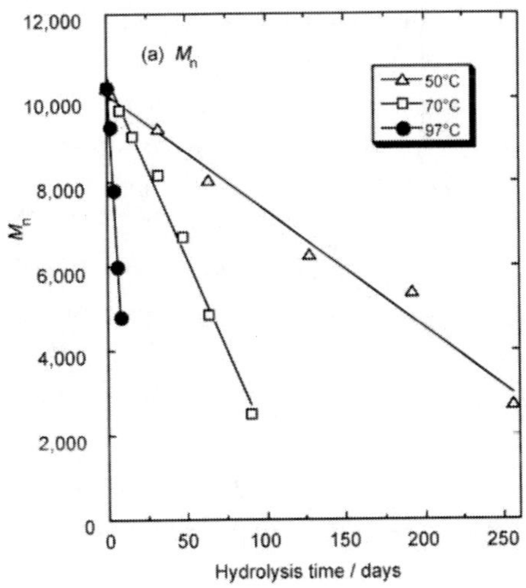

Figure 4. Mass-averaged molecular mass (M_n) of PLLA crystalline residues (extended-chain crystallites) hydrolyzed at different temperatures as a function of hydrolysis time. Reprinted with permission from In vitro hydrolysis of poly(L-lactide) crystalline residues as extended-chain crystallites: II. Effects of hydrolysis temperature, Hideto Tsuji, Kensaku Ikarashi, Biomacromolecules 2004, 5, 1021-1028. © (2004) American Chemical Society.

PLA can be endogenously degraded by serine proteases, such as proteinase K and elastase; L-PLA is slower degraded than (D,L)-PLA.

Thermal Degradation

Thermal degradation takes place *via* breakdown of the chain and depolymerization, again leading to cyclic and linear oligomers and their diastereoisomers. The reactions involved are usually oxidative degradation, *cis*-elimination, and intra- and exo-molecular transesterification. This process usually involves random breakdowns forming cyclic and linear oligomers which can be recombined, specially at temperatures below the boiling point of lactides around 250 °C [32-34]. From the mass loss along the degradation process, it has been found that cyclic lactide and other species, such as CO_2, CO and acetaldehyde are evolved [35], total pyrolysis taking place at 240-270 °C.

The presence of polymerization catalysts containing Sn, Zn, or Al delays the degradation to higher temperatures, pyrolysis taking place at 280-370 °C [36]. Tin compounds promote degradation *via* intramolecular transesterification [37]. Several authors [38] have proposed that stability against thermal degradation in the presence of metal salts decreases in the order Fe>Al>Zn>Sn; so a precise selection of the polymerization catalyst is required, as its residues can contribute to degradation. Abe et al. [39] have studied the catalytic effect of residues of Zn compounds originally used as polymerization catalysts, such as diethyl zinc and zinc didodecanoxide; their effect is similar to that found for Sn catalysts, but depends on the zinc atom content; zinc compounds usually catalyze the intramolecular transesterification, forming linear oligomers and selective depolymerization, forming lactides.

Encapsulation

PLA is biocompatible (it is not rejected by the living organism), biodegradable and immunologically inert (it does not provoke allergy), atoxic and completely reabsorbable. PLA micro and nanoparticles have been very much studied for encapsulation and release of a large number of therapeutical molecules [40-42]. Nanoparticles are also interesting to improve the stability of the encapsulated drug and for delivery and controlled release of drugs to different parts of the organism.

Organic/inorganic composites integrated in polymers constitute a sort of synergic system, with better properties and performances than the traditional composites and pure polymers. In addition, such an improvement in the properties takes place without a noticeable increase in the density, nor a decrease in their optical and reciclability properties. Such a behaviour is a result of the low loading required to obtain an even large increase in the mechanical properties (together with their thermal stability and resistance to fire). Permeability against gases and liquids is also reduced and when using conducting or biodegradable polymers, these properties are also improved [43].

The most widely used polymers to prepare these organic/inorganic/polymer systems are polylactic acid, PLA, and its copolymers with glycolic acid (PLGA), as they are also biodegradable, biocompatible and inocuous [44]. Other biodegradable polymers with promising properties along this line are other polyesters, like policaprolactone [45] and polyanhydrides [46].

An interesting property of these polymers is that their ability to release the active molecules depends on their degradation profile. Consequently, if we are able to control the degradation of the polymer, we will be able to control the release of the active molecule. It is well known that degradation of polyesters leads to formation of acid oligomers which can be accumulated within the particles, enhancing the strength of the acidic medium of the polymer network and decreasing the pH inside the particles.

Copolymers: PLGA

The successful use of copolymers within a living organism depends on the chemical structure and distribution of the monomer units along the polymer chain. Nanoparticles with a hydrophobic surfaces (e. g., PLA and PLGA) are rapidly assimilated by the cells of reticuloendothelial systems [47]. Copolymers can be designed to increase their surface hydrophylic character, in order to extend the time available for drug delivery. PLGA is widely used in biomedical applications because it is hydrolyzed in water, forming its monomers (lactic acid, LA, and glycolic acid, GA), two products that participate in metabolic reactions in the organism and do not give rise to adverse reactions.

Synthesis of PLGA

Low molecular mass PLGA copolymers can be prepared by condensation of lactic acid and glycolic acid removing water by vacuum treatment. The reaction is carried out using diphenylether as solvent and Sn catalysts. The high molecular mass polymers are prepared by ring opening (ROP) of lactide with glycolide *in vacuo* using THF, toluene or dioxane. Although the optimum reaction temperature is 100-150 ºC, it is usually carried out at lower temperatures to avoid inter- and intra-molecular detransesterification processes. The ring opening process (ROP) is favoured in the presence of Zn-containing catalysts, such as zinc lactate, and of Sn^{II} compounds. By controlling the reaction temperature and the catalyst used, the relative amounts of lactic and glycolic monomers in PLGA can be controlled, and thus its hydrophobicity; for instance, using Sn^{II} catalysts at 160 ºC the polymer obtained contains 70 % lactic acid [48]. Block copolymers can be prepared by sequential addition of monomers or by inserting the hydroxyl end of the chain in a homopolymer [49].

Degradation of PLGA

Hydrolytic degradation of PLGA takes place by selective breakdown of the ester bond connecting glycolic units and removal of the glycolic units, thus giving rise to a decrease in the glycolic acid content. The composition of the polymer affects the degradation process and also the crystallinity. When the glycolic content is increased from 15 % to 50 % the degradation time is reduced from 5-6 months to 1-2 months [50]. The polymers with 25-70 % glycolic acid are essentially amorphous and are used as supports to release drugs because of their easy biodegradation.

Layered Double Hydroxides (LDH)

Layered double hydroxides (LDH) are a large group of natural and man-made compounds first identified in Scandinavia around the middle of the XIX century. They are formed by hexagonal close packed $[Mg(OH)_6]$ octahedra sharing edges to form layers (brucite), with a certain Mg^{II}/Al^{III} isomorphical substitution; the resulting positive charge is balanced by hydrated anions in the interlayer space, where water molecules also exist. A scheme of the structure is outlined in Figure 5. X-ray absorption spectroscopy (XAS) studies have demonstrated that only octahedral holes between the hydroxyl layers are occupied, and all the tetrahedral holes are empty [51]. Even in the case of compounds containing Zn^{II} cations substituting the Mg^{II} ones, prone to form tetrahedrally coordinated compounds, EXAFS/XANES studies have shown the presence of $[Zn(OH)_6]$ units, similar to the hexa-aquo-cation, $[Zn(H_2O)_6]^{2+}$ [52].

The most representative member of this family is *hydrotalcite*, with the formula $[Mg_6Al_2(OH)_{16}]CO_3 \cdot 4H_2O$. Many compounds with this structure are known, but with other cations (Mg, Mn, Fe, Co, Ni, Cu, Zn, …, and Al, V, Cr, Fe, Ga, Mn, Y, …) in the brucite-like layers and other anions (chloride, nitrate, sulphate, phosphate, …, oxometalates, polyoxometalates, Keggin-type anions, anionic coordination compounds, organic anions, surfactants or biological anionic moieties) in the interlayer space [53]. These compounds are known as *hydrotalcite-like* compounds.

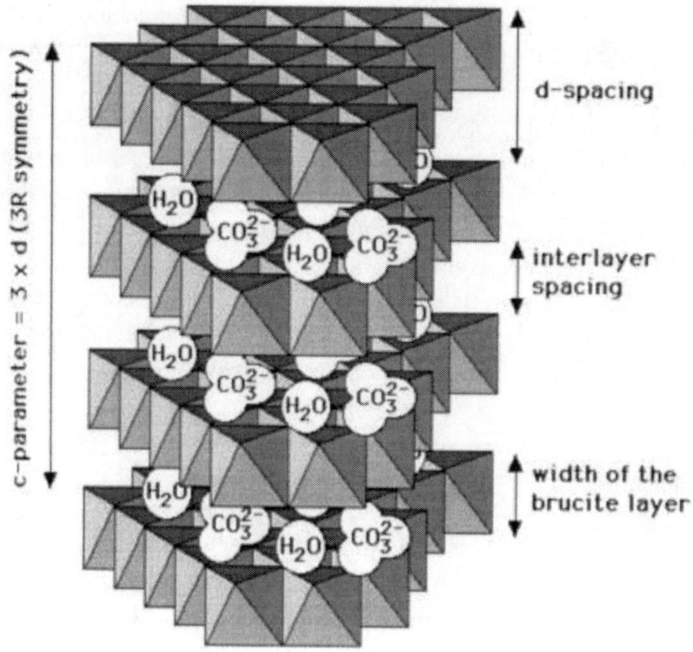

Figure 5. Stacking of brucite-like layers forming the hydrotalcite structure. Reprinted from Coordination Chemistry Reviews, Vol. 181, Vicente Rives and Maria Angeles Ulibarri, Layered double hydroxides (LDH) intercalated with metal coordination compounds and oxometalates, pp. 61 – 120, © 1999, with permission from Elsevier.

The chemical composition of LDHs can be written as $[M^{2+}_{1-x}M^{3+}_{x}(OH)_2]\ (A^{n-})_{x/n} \cdot m\ H_2O$. The value of x, the molar Al/(Mg+Al) ratio, should be within a rather narrow range ($0.20 \leq x \leq 0.33$) to obtain systems free of contamination with the hydroxides of the divalent cation (for very low x values) or the trivalent one (for very high x values). The layers can be stuck in different geometries [54], the commonest one being rhombohedral, where the unit cell along the stacking direction corresponds to three brucite-like layers.

The interlayer anions are exchangeable and consequently these compounds are also known as *anionic clays* [5]. Research on preparation, characterization and applications of hydrotalcite-like compounds has increased sharply in recent years [55-60].

Structure

For a rhombohedral stacking of the layers, the first diffraction maximum recorded at low diffraction angles corresponds to planes (003), and usually two or even more harmonics are recorded. The maxima recorded in the middle range of the 2θ axis are usually broad and asymmetric as a consequence of turbostratic disorder [61]. The doublet recorded around 60 ° (2θ) is highly informative, and is usually recorded as a single maximum when the intercalated anions are organic ones. The first of these two maxima corresponds to diffraction by planes (110) and its spacing corresponds to one half of the average cation-cation distance within a brucite-like layer, while the value of parameter *c* corresponds to three times the spacing corresponding to diffraction by planes (003).

The interlayer space and thus the value of parameter *c* is very sensitive to the hydration degree and to the molecular size and orientation of the interlayer anion, insertion of bulk anions requiring swelling of the layers. The orientation of non-spherical anions (for instance, flat anions, such as nitrate), with their molecular plane parallel, perpendicular or tilted with respect to the brucite-like layers also accounts for changes in the height of the interlayer region.

Although the hydrotalcite-like crystals may show different morphologies, the most frequently found corresponds to think layers with a hexagonal shape.

Any of the different anions existing in the medium where the LDH is being prepared can be intercalated, although a sequence of preferential intercalation has been found; carbonate is very easily intercalated, and the anions most easily removable are nitrate and chloride.

Thermal Properties

Thermal decomposition of hydrotalcites takes usually place in three consecutive steps, although the whole process depends on many different factors, such as (i) the specific nature of the layer cations, and of the interlayer anions, and (ii) the nature of the atmosphere (oxidant, inert or reductive) surrounding the sample during the decomposition process. Concerning the layer cations, differences are found between those able to undergo oxidation process (e. g., Co^{II}, Mn^{II}, V^{III}, Cr^{III}, Fe^{II}, etc.) and those which do not (Mg^{II}, Zn^{II}, Al^{III}, Ga^{III}, Fe^{III}, Ni^{II}, etc.), Concerning the anions, they can undergo oxidation reactions, and in addition they can evolve to gaseous, non permanent species upon calcination (carbonate, nitrate, etc.), or can be permanent species, i. e., anions which are transformed into metal (from the layers) salts upon decomposition, as in the case of oxo and polyoxometalates.

The most widely studied system has been again MgAl-CO3. A smooth mass loss takes place below 150 °C (the corresponding endothermic effect in the differential thermal analysis (DTA) curve is usually ill-defined), due to removal of water molecules, weakly adsorbed on the external surface of the crystallites. As the temperature is increased, a much more important mass loss amounting up to 12-15 % of the initial sample mass corresponds to removal of interlayer water molecules; in the DTA trace it is recorded as a rather symmetric and intense endothermic effect. If the thermogravimetric (TG) effect is sufficiently isolated from others, the amount of interlayer water molecules can be straightforward calculated, and, together with element chemical analysis data for the metallic elements, the formula of the solid can be concluded.

Removal of interlayer water molecules gives rise to a decrease of the interlayer spacing, as concluded from the observed shift of the maxima due to diffraction by basal planes towards larger diffraction angles [62].

The most important mass loss (even up to 30 % of the original mass) is usually recorded between 200 and 550 ºC, and is due to removal of hydroxyl groups from the brucite-like layers as water, together with removal of the interlayer anion in the form of volatile species (e. g., CO_2 from carbonate, or NO_x from nitrate). The DTA curve shows usually a broad, non-symmetric, endothermic effect, usually including both processes, which are only distinguishable from the DTG curves [63]. Mass changes are not usually observed above 550-600 ºC, although in some cases carbonate elimination is not completed even up to 800-900 ºC [64]. The DTA effects recorded above 600 ºC generally correspond to crystallization of mixed oxides and red-ox processes.

Powder X-ray diffraction (PXRD) monitoring of the crystalline phases formed and mass spectrometric analysis of the gases evolved during decomposition [63,65,66] results suggest the following steps to account for the decomposition of hydrotalcite (equations are not balanced):

$$[Mg_{1-x}Al_x(OH)_2]\,[(CO_3)_{x/2} \cdot n\,H_2O] \rightarrow [Mg_{1-x}Al_x(OH)_2]\,[(CO_3)_{x/2}] + n\,H_2O$$
$$[Mg_{1-x}Al_x(OH)_2]\,[(CO_3)_{x/2}] \rightarrow Mg\,(Al_x)\,O_{(2+3x)/2} + CO_2$$
$$Mg\,(Al_x)\,O_{(2+3x)/2} \rightarrow MgO + MgAl_2O_4$$

Preparation of LDHs

Many methods have been described in the literature to prepare LDHs; by using one or another the properties of the final solid (and its suitability for a given application) can be predetermined. Only those most commonly used for the purposes here depicted are here described.

Precipitation

They are the most widely used ones. The properties of the final solids depend on the specific precipitation process followed, the nature of the reagents, their concentrations, precipitation pH, temperature, ageing time (if any) after precipitation, the presence of impurities, and the washing and drying conditions [67].

Precipitation at decreasing pH. A solution containing the metal cations is added to a basic solution of an alkaline metal (usually NaOH or KOH). The crystallinity of the resulting solids is rather low and they need to be submitted to an ageing treatment to improve it [68].

Precipitation at increasing pH. It consists of titration of the solution containing salts of the metal cations with a basic solution, usually an alkaline metal hydroxide or carbonate [69,70]. Despite a sequential precipitation of the metal hydroxides would be expected, it seems that precipitation of $M(OH)_3$ in some sort of way induces the simultaneous precipitation of $M(OH)_2$.

Precipitation at constant pH (coprecipitation). It consists of the slow addition of a solution of the metal cations to water or to a solution of the anion to be intercalated. A basic solution is simultaneously added to keep a constant pH value, leading to the simultaneous precipitation of the hydroxides. The process is based on the condensation of aquocomplexes in solution, eventually leading to formation of brucite-like layers with both cations homogeneously distributed in the layers, and with intercalated anions. Experimental studies have shown that the layers and the interlayer domains are immediately formed, without needing delamination of the brucite-like layers [71].

The Salt-Oxide Method

A suspension of the oxide of the divalent cation is first prepared, and vigorously stirred in an aqueous solution of the trivalent metal cation in excess above the pursued stoichiometry of the LDH, during several days at room temperature [72]. This method was initially used to prepare Zn,Cr-Cl LDHs [73], but it has been also successfully applied to prepare LDHs with other layer cations and interlayer anions.

Anion Exchange

The layered nature of LDHs permits their easy stacking, the interlayer anions being able to migrate rather easily through the interlayer space and can be topotactically exchanged [74]. This ability has been very often used to prepare new LDHs, with different anions, from other LDHs. Using bulk organic anions as exchange species facilitates the exfoliation of the hydroxyl layers [75] to prepare nanocomposites [76]. The following sequence in the selectivity of the layers for monovalent and divalent anions has been found [77]:

$$OH^- > F^- > Cl^- > Br^- > NO_3^- > I^-$$
$$CO_3^{2-} > C_{10}H_4N_2O_8S^{2-} > SO_4^{2-}$$

Due to the electrostatic interactions, selectivity for divalent anions is larger than for monovalent ones and thus LDHs with nitrate or chloride are rather suitable to be used as reagents for anion exchange processes under vigorous agitation or stirring of a suspension of the LDH in a solution of the anion to be inserted. The reaction pH should not affect both the initial and the final solids, nor the nature of the anion to be inserted. Ultrasounds have been used sometimes to accelerate the exchange process [78].

The Reconstruction Method

This method is based on the so-called *memory effect*, i. e., the ability of mixed oxides formed by gentle calcination (usually not exceeding 500 – 550 ºC) of LDHs with intercalated carbonate, to *recover* the original layered structure, inserting anions from the medium [79,80]. The mixed oxides are also able to recover the layered structure by simply standing in open air, intercalating hydroxyl or carbonate groups from atmospheric water vapour or carbon dioxide, respectively [81,82].

The Sol-Gel Method

The first step is to prepare a sol by addition of a metal alkoxide to a solvent, with which it reacts, forming the colloidal suspension. Ageing leads to the gel, which calcination leads to formation of the final solid [83]. The textural and morphological properties both of the layered materials prepared by this method as well as of the oxides obtained upon their calcination, are different from those shown by the same compounds, but prepared by classical precipitation procedures [84,85].

Urea Hydroysis

Strictly speaking this is a constant pH method. An excess of urea is added to the aqueous solution of the metal cations and the mixture is heated [86]. Upon decomposition, urea gives rise to ammonium and carbonate ions, the pH increasing homogeneously leading to a so-called *homogeneous* precipitation of the LDH, with inserted carbonate anions [87,88].

Post-Synthesis Treatments

Preparation variables are very important in determining the physical properties of LDHs; in some cases, poorly crystallized materials are obtained and thus post-synthesis treatments need to be applied.

Hydrothermal Treatment

When a suspension containing the recently precipitated LDH is submitted to hydrothermal treatment (i) the hydrotalcite-type phase is formed; (ii) the small crystallites existing in the suspension are converted into larger crystals (Ostwald ripening), which permits in some cases an easier and more reliable characterization; and (iii) amorphous precipitates crystallize. Two different approaches can be applied: (i) hydrothermal treatment above 100 ºC in an autoclave, aiming to accelerate the reaction, or (ii) below 100 ºC; in this case the process is usually known as *ageing*. Hydrothermal treatment enhances the crystallinity of the solid, and then their PXRD diagrams show sharper and more intense diffraction maxima. The crystallite size increases, and also the particle size, decreasing the specific surface area. Additionally, the FTIR spectra show sharper and more intense absorption bands, due to formation of better ordered, undistorted, solids [89].

Microwave Treatment

The main advantage of using microwave (MW) irradiation as an energy source during hydrothermal treatment is that the crystallinity degree reached is of the same order as with hydrothermal treatment with conventional (furnace) heating, but the solids show a smaller particle size (and then a larger specific surface area), in a shorter period of time and saving in energy consumption [90]. Microwaves are electromagnetic radiations with wavelength between 1 cm and 1 m (corresponding to a frequency between ca. 300 MHz and 300 GHz); they are highly coherent and polarized. Those with a frequency of 2.45 GHz (wavelength 12.2 cm, equivalent to $1.02 \cdot 10^{-5}$ eV) [91] are used for commercial and chemical applications. Not all materials are sensitive to MW, and thus the materials can be classified in conducting, insulators and absorbents [92].

When using MW radiation, energy is directly transferred by interaction with the electromagnetic field and is transformed into thermal energy; it is more convenient to describe the process as *energy conversion* rather than *energy transfer* [93]. MW will penetrate the materials, heating taking place on the bulk of the material (*volumetric heating*) and results faster and uniform, even for dense materials. Substances interact differently with MW radiation: Liquids or solids cannot rotate completely free and thus their heating is not a result of absorption of MW involving transitions between rotational energy states (as for gases). Electrons in a conducting material move freely as a response to an externally applied electric field, giving rise to an electric current; unless the material is a superconductor, heating arises from the resistance to the movement of the electrons (*Joule* or *conducting heating*). Most of metals reflect MW and they will not become heated. Electrons are not free to move in insulators, but electron reorientation or distortion of permanent or induced dipoles may produce heating, known as *dielectric heating* [94]. In ionic materials where the ions can move under an applied field, a *Joule heating* may arise because of the ionic current. Electric dipoles in solids do not reorientate instantly, but possess a typical *retarding time*, τ; if the electric field of the applied radiation changes very rapidly (t is very small, $t \approx \tau$ or $t < \tau$), the dipoles are unable to reorientate at the same speed and heat arises.

The main advantages of MW heating are [95]: (i) higher heating rate, (ii) lack of direct contact between the heating source and the materials, (iii) selective, controlled heating, and (iv) smaller equipment and energy saving. As a result, MW have been used as a radiation source in microwave spectroscopy and electron spin resonance; they are also used in Analytic Chemistry to digest samples [96], drying, sample cleaning, etc. Also because of selective heating and faster reaction rate, they have been used in organic synthesis [97]; in Environmental Engineering [95] they have been used for waste treatment, recycling, etc. [98]. They have been used in Solid State Chemistry to join ceramics [99], to synthetize compounds in the absence of solvents, substituting conventional heating in ceramic processes [83], to prepare pillared clays [100] and LDHs [101-104]. They also find application in Catalysis, both to activate catalysts and to increase the reaction rate [105].

Applications of LDHs

LDHs and the oxides formed upon their calcination show many industrial applications; in addition, pillaring or intercalation leads to formation of tridimensional structures, opening new routes to prepare solids with uncommon properties.

LDHs are used as anion exchangers and adsorbents [71,77,106]. Their exchange capacity and resistance to heating are larger than those of conventional resins. Exchange is strongly difficult when starting from carbonate-containing phases, and thus it is usually carried out starting from the nitrate or chloride forms. pH is also important, as in some cases exchange ought to be carried out at pH values for which the LDH structure is unstable. The so-called *memory effect* [79,80] and their reciclability makes the LDHs highly suitable for anion scavenging: in this way, species like arsenate, chromate, selenite or vanadate, as well as organic pollulants, such as phenols [107-109], or plaguicides [110] have been successfully removed from natural water [105,111].

LDHs are used in Heterogeneous Catalysis well as catalysts, catalyst precursors or catalyst supports, due to the optimum properties shown by the solids formed upon controlled thermal decomposition of the LDHs: (i) large specific surface area (usually in the range of $100 - 300$ m^2 g^{-1}); (ii) homogeneously dispersed metal cations, which upon reduction lead to small stable metal crystallites; (iii) synergic effects between the metal ion components, developing some unusual basic or hydrogenating properties; and (iv) memory effect, permitting reconstruction of the layered structure under mild conditions.

Their use in several heterogeneous catalytic processes has been reviewed [112-116]. Important applications have been found in olefin polymerization, aldol condensation, hydrocarbon reforming, methanol synthesis, Fischer-Tropsch synthesis, and oxidation reactions [117-119]. The mixed oxides formed upon their calcination are active in hydrogenation reactions [120-124].

Pillared layered double hydroxides behave as shape-selective heterogeneous catalysts, as well as to stabilized homogeneous or biomimetic catalysts, able to extend their active life and facilitating their recovering and recycling; they have been also used to prepare supported catalysts with larger loadings of active phase and larger stability and activity than those prepared by conventional methods [125].

In Medicine and Pharmacology they have been used as antacids [126] and to host and control delivery of drugs [9,10,127-129]. This ability has been also useful for controlled delivery of fertilizers [130], to prepare dyes for controlled dyeing [131], etc.

The mixed oxides formed upon their thermal decomposition (also from non-stoichiometric LDHs) have been used as ceramic pigments [132-139]; the colour can be tuned by modifying the precise nature of the transition metal cations in the brucite-like layers and also by fine tuning of their relative concentrations.

One of the applications with a large industrial impact has been when added to poly (vinyl chloride), resulting in its strengthening and longer duration of the white colour [140,141]; as combustion retardants [142,143] and heating resistant insulating wires [144]. Their cations can be tuned to absorb infrared and ultraviolet radiation, making them suitable to prepare protecting films [145-147], etc. Due to their potentiometric and colorimetric properties they can be used as sensors in detection devices [148]. Their structural anisotropy makes them suitable to prepare transparent layers and membranes [149].

Polylactic Acid As a Drug Support, Carrier, and Delivery Agent

Cam and Marucci [38] have studied the thermal stability of PLA and the effect of the presence of metal ions on its degradation. High (700 kDa) molecular mass PLA decomposed leading to formation of smaller molecular mass residues, monomers and oligomers (lactic acid and lactides), which did not affect degradation of the polymer, but the presence of metal ions (Zn, Fe, Sn, …) modified the degradation pattern, as they accelerated the intra- and inter-transesterification reaction at high temperature (around and above 240 ºC). The high degrading power of iron seemed to be related to its ability to coordinate ester groups, accelerating transesterification and depolymerization.

Fan et al. [150] have prepared PLA (molecular mass 158 kDa) to which CaO or MgO were added (5 % w/w) with vigorous stirring to disperse the inorganic particles. Below 250 ºC, the presence of calcium led to decomposition of the polymer and racemization of the residues; however, the presence of MgO delayed the degradation. Above 250 ºC almost complete (98 %) decomposition followed the same patterns both in the presence of CaO or MgO, although formation of lactic acid in the presence of MgO would permit its recycling.

Park and Xanthos [151] have studied the thermal and hydrolytic degradation of PLA by ionic liquids with tetrafluoroborate or decanoate. The results indicated that the thermal degradation was faster with decanoate, but the hydrolytic degradation was faster with tetrafluoroborate. Hydrolytic breakdown of the ester bond was catalyzed by the carboxylic groups of PLA.

Dagnon et al. [152] have reported release of ibuprofen (IBU) from IBU/PLA mixtures (1% w/w) in the form of films prepared in a custom-built spin caster without the need for any substrate. The drug release was related to the thermal degradation of the polymer, and proceeded by transesterification, leading to formation of lactic acid and cyclic oligomers which underwent formation of acrylic acid oligomers and bonds breakdown, to acetaldehyde and carbon dioxide. Thermal degradation of pure PLA started at 285 ºC and was completed at 430 ºC. The presence of IBU changed the degradation process and decreased the thermal stability of the polymer. When incorporating LDH-loaded IBU to the polymer the degradation temperature of the polymer also decreases (the effect of LDHs will be studied below). The DTG minimum for PLA was recorded at 387 ºC, decreasing to 331, 317, and 311 ºC for LDH loadings of 1, 3, and 5 % (w/w), respectively.

Miao et al. [153] have incorporated IBU into PLA, forming fibers by electrospinning; the thermal degradation and the release of the drug have been studied. Probably due to the low IBU loading (2 % w/w), and contrary to the results above mentioned by Dagnon et al. [152], these authors did not observe any change in the temperature at which degradation of the polymer started upon addition of IBU; the drug seemed to be well dispersed in the amorphous polymer matrix, as no diffraction maximum of the drug was recorded in the PXRD diagram of the composite. The amount of IBU released after 2 h was less than 10 % of the total IBU loading; this fact has been related by these authors to the low flexibility of PLA fibers, impeding the mobility and diffusion of water, which is responsible for PLA degradation, a necessary step previous to release of the drug.

DeLeon et al. [154] have prepared IBU/PLA samples with 1.2 and 2 % (w/w) loading by spin casting; these samples were used as reference materials to study the release of IBU from LDH-IBU systems dispersed in PLA. The PXRD diagrams showed maxima due to the PLA matrix, a broad maximum and two sharper ones due to the semicrystalline polymer. IBU release was followed by HPLC. IBU/PLA films showed an important release during the first 5 h, followed by a much slower release until a flat profile was reached. This behaviour was probably due to release of IBU which was originally adsorbed on the surface of the polymer units. On studying the diffusion mechanism [155,156] it was concluded that after the first release steps from the surface of the PLA particles, release from deeper sites was difficult and very slow.

Zhang et al. [157] have prepared microspheres containing PLA, IBU and Fe_3O_4; the components (500:100:40 mass ratio) were suspended in dichloromethane and submitted to ultrasounds (100 W) for 5 min; the emulsion formed was stirred for 3 h at room temperature until the solvent was completely evaporated, and the dry powder was liophilized for 48 h. Another sample was similarly prepared, but without Fe_3O_4. In all cases the IBU and magnetite particles were retained within the microsphere cavity (diameter 1.6±0.8 µm), Figure 6. Thermal decomposition of PLA started at 200 ºC and no further mass loss was observed above 450 ºC, indicating that IBU had been also removed, remaining Fe_3O_4 exclusively.

IBU release studies were carried out at pH 1.2 and 7.4. The process was faster and the amount of IBU larger at pH 7.4 (73.08 %) that at pH 1.2 (41.04 %). In both cases, a fast release (extending for ca. 3 h) due to removal of IBU from the surface of sites close to the PLA surface [158] was followed by a slow release extending for 50 h. The presence of Fe_3O_4 somewhat delayed the release process. This depended on the solubility of the drug and the permeability of the polymer at the pH the experiment was carried out. The drug can be released by dissolution, diffusion or degradation of the polymer [159]; diffusion of IBU from inside the microspheres could be impeded in the presence of magnetite, and thus, the release process could be probably modulated by using different amounts of magnetite.

Hu et al. [160] have also prepared IBU/PLA systems (9% w/w) and have monitored the drug release by spectroscopy at 264 nm. The fibrous PLA membranes were prepared by electrospinning. At 11 kV with a 0.04 ml min^{-1} flow of an ethanol solution containing PLA and IBU. Electron microscopy (scanning, SEM, and transmission, TEM) results evidenced that IBU was encapsulated within the PLA fibers. The sample was incubated in physiological serum with phosphate buffer solution (PBS) buffer for 100 days at 37 ºC. The accumulated amount of drug released was 94 % of the initial amount of IBU after 20 days; 64 % was released during the first 12 h. When the sample was modified by adding mesoporous silica the release was even slower: one hundred days were needed to release 91 % of the drug and during the first 12 h only 6 % was released, in both cases lower values than from the IBU/PLA system. *In vivo* studies demonstrated that the slow release of IBU improved its antiinflammatory action and prevented formation of peritendinous adhesions.

Zhou et al. [161] have prepared samples containing indomethacin (IND) and other antiinflammatory non steroideal drugs (NSAID) dispersed in PLA. Their release has been related to the structure of the polymer. The corresponding drug and PLA were first dissolved in THF at 60 ºC and the solution was cooled to 20 or -75 ºC for 12 h; after washing with distilled water the samples were dried and liofilized. The drug/PLA loadings were in the 0.1:1 to 2:1 range.

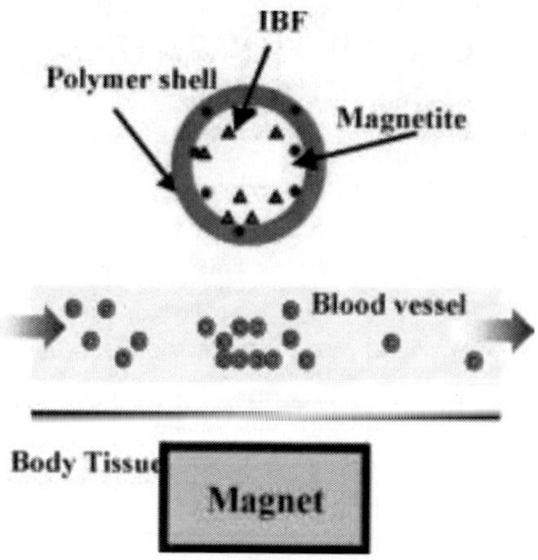

Figure 6. Scheme describing the effect of magnetite particles within polymer nanospheres containing ibuprofen to facilitate their flow in a blood vesel. Reprinted from Colloids and Surfaces A: Physicochemical and Engineering Aspects, Vol. 432, Xiangyu Zhang, Li Xue, Jun Wang, Qi Liu, Jingyuan Liu, Zan Gao and Wanlu Yang, Effect of surface modification on the properties of magnetic nanoparticles/PLA composite drug carriers and in vitro controlled release study, pp. 80-86, © 2013, with permission from Elsevier.

The samples prepared at 25 ºC were constituted by nanolayers, changing to nanofibers for loadings of and higher than 5 % IND. The quality of the nanofibers was improved when larger amounts of drug were loaded, as crystallization of IND within the matrix favoured formation of the PLA nanofibers. However, when the samples were prepared at -75 ºC nanofibers were formed in all cases, whichever the IND loading, but the diameter of the nanofibers decreased as the drug loading was increased; probably, the larger IND content in some sort of way favoured an increase in the number of crystallization sites where the polymer chains were formed. The IND crystals were uniformly distributed in the PLA matrix (nanolayers or nanofibers).

The release of IND was monitored by UV-Vis spectroscopy at 265 nm. A first, burst, release was observed, amounting 22 % of the initial amount of IND in the sample prepared at 25 °C, and 40 % for that prepared at -75 °C; release from the sample prepared at 25 °C extended for a longer period of time (months). These authors concluded that diffusion of water molecules in the nanofibers was easier than in the nanolayers, favouring interaction with the IND crystals and accounting for its easier release.

Montes et al. [162] have prepared samples with naproxene (NAP) dispersed in PLA and Eudragit® (EUD) using ethanol as a solvent and CO_2 under supercritical conditions. CO_2 was pumped under supercritical conditions on the ethanolic solution containing the drug and the polymer provoking a supersaturation and leading to precipitation of the drug/polymer compound as a powder. Spherical particles with a broad particle size distribution were formed in all cases; their size was always larger (ca. 10 times) for PLA than for the EUD system. The PXRD diagrams showed sharp maxima due to crystalline NAP and a broad maximum associated to the polymer; the presence of NAP was further confirmed by FTIR spectroscopy.

The drug loadings were from 1:1 to 1:5 (NAP:PLA) ratios. The release was followed by UV-Vis spectroscopy at 273 nm and two experimental conditions were used: in simulated gastric fluid (SGF, pH=1.2) and in simulated intestine fluid (SIF, pH=6.8). Release was initially fast, probably corresponding to surface adsorbed drug molecules. The release of EUD in SGF was slow, due to the lack of degradation of EUD at pH 1.2; however, the amount of NAP released under SIF conditions was larger from EUD than from PLA, as at pH 6.8 EUD was dissolved, while PLA was degraded by diffusion of water. The release rate of NAP did not depend on the initial drug loading.

Following previous studies on release of phenol red from PLGA [163], Onishi et al. [164] have prepared tables with KET supported on poly(DL-lactic acid-co-glycolic acid) copolymer, PLGA, with a DL-lactic acid/glycolic acid ratio of 1:1 (mol/mol); its release *in vivo* and *in vitro* was studied. The tablets (30 mg, 3.2 mm width) were prepared by direct compression at room temperature, containing 1 or 5 mg of drug at room temperature. The release profiles *in vitro* in 100 ml PBS at pH 7.4 were similar to each other whichever the amount of drug/polymer composite, and thus the *in vivo* studies with rats were carried out using only tablets containing 5 mg of KET. The first steps of the release were faster *in vivo* than *in vitro*, but the opposite in the final stages, the drug levels being below the recommended therapeutical dose. It is expected that the process can be optimized by using different ratios of PLA and PLGA with different molecular masses.

Corrigan and Li [165] have studied the mechanism governing the release of small molecules (KET, IND and coumarin-6) and macromolecules (human serum, albumin, ovalbumin) from 50:50 PLGA; the samples had been prepared by emulsion/evaporation and the release experiments were carried out in PBS (pH 7.4) at 37 °C. The release curves showed a first burst stage where the amount of drug released depended on the specific drug studied, its loading and its solubility; it was higher when higher drug loadings were used or it was more soluble. The drug was released from the surface of the polymer, which seemed to act merely as a support. This fast stage was followed by a slower one where the drug molecules originally trapped in the PLGA matrix were released; the process depended on degradation of the polymer; as these drugs are acidic they accelerated degradation of PLGA and their release was faster than those of the proteins studied.

Zakeri-Milani et al. [166] prepared a vancomycin (VAN)/PLGA composite aiming to maintain high concentrations of the drug through the intestine and thus to improve the performance of this antibiotic when orally administrated. Different amounts of VAN were incorporated into 50:50 PLGA using the double-emulsion solvent evaporation method. Depending on the specific composition of the composite the size of the nanoparticles ranged between 450 and 466 nm, and the PXRD diagrams did not show any significant difference among the different samples, suggesting that no crystallization of VAN has taken place. No difference could be observed either in the FTIR spectra, suggesting that the chemical interaction betwen VAN and the polymer was minimal, if any. Release took place in two consecutive stages, a burst, first one, releasing between 10.78 and 12.27 % of the initial amount of drug in 30 min (VAN molecules in or near the surface of the polymer particles), followed by a second, slower stage, adjustable to an exponential model extending for 6-8 h following a first order kinetics, attrributed to diffusion of the drug inserted within the PLGA structure. The release rate was not dependent on the polymer content and all preparations exhibited similar dissolution efficiencies. *In vivo* essays with rats showed that intestine permeability for VAN was improved when supported on PLGA, which opened the possibility of easier formulations than those currently available for this drug.

Hami et al. [167] have prepared docetaxel (DOC)-conjugated PLA-(polyethylene) micelles; this drug is used to treat several types of cancer. The composite was able to be directed to folate receptors, where the drug was released in a pH-dependent process, and the drug being covered with the polymer made the process advantageous, as secondary effects were decreased. The TEM micrographs showed spherical micelles with a size below 200 nm. The release of the drug was studied by the dialysis method in phosphate (pH 7.4) and acetate (pH 5.0) buffers to mimetize the physiological and acid conditions of the tumoral tissue, respectively. As much as twice of the drug was released at pH 5.0 if compared to that released at pH 7.5, thus suggesting that the process might be usewful to kill tumoral cells.

Bédouet et al. [168] tried to cancel the first burst stage usually observed in the release of drugs from drug/polymer composite, as this first step led to release of a large amount of drug which is misused by the organism. They have synthetized composites of IBU with degradable hydrophilic microspheres with oligo(ethylene-glycol) methacrylate and poly (PLGA) dimethacrylate.

The release studies showed a time maintained, slow release, probably due to formation of a polymer-drug covalent bond *via* an ester bond; the released drug molecules maintained their antiimflammatory activity. The release of the drug and polymer hydrolysis took place simultaneously, and so it is necessary to optimize the process to diminish the degradation products from the microspheres.

Zhang et al. [169] have studied the release of doxorubicine (DOX) and curcumin from micelles contained in copolymers. When using PEG-PLGA (PEG=polyethyleneglycol) the release of the drug molecules which were physically trapped within the polymer matrix took place by diffusion; however, when chemical interaction between the drug and the polymer existed the release proceeded well by migration of the micelles outside the copolymer matrix or by destruction of the polymer and release of the drug from the portion of polymer to which it was bonded [170,171].

Release of the drug depended on the degradation of the copolymers induced by the presence of water. It has been mentioned that the degradation easiness depended on the molecular mass of the polymer [172], the length of the hydrophylic chains and the amorphous character of PLGA. Degradation was faster for low molecular mass polymers than for high molecular mass ones. When a chemical interaction existed degradation took place by hydrolysis of the ester group, and the DOX/PLGA oligomers were slowly released along time. When the ester bond was broken, -COOH and –OH groups were formed, and the former favoured further breaking of ester bonds by a self-catalyzed process [173]. The overall mechanism could be understood as a combination of diffusion of soluble oligomers and neutralization of the carboxylic group [172]. Final degradation products were LA, GA, and LA and GA oligomers.

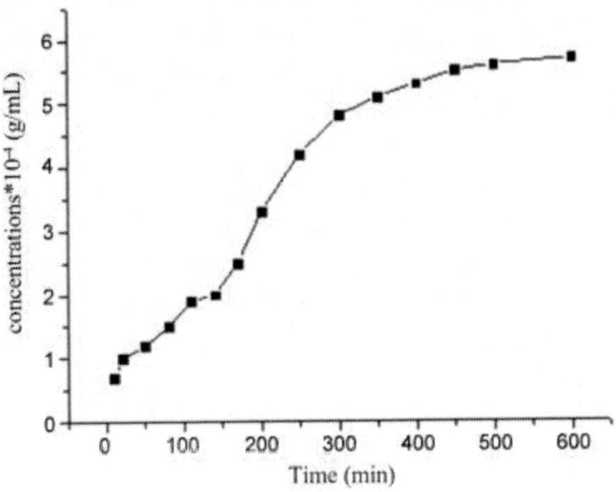

Figure 7. Release profile of PLGA-coated nano-curcumin. Reprinted from The Journal of Supercritical Fluids, Vol. 89, Fatemeh Zabihi, Na Xin, Sining Li, Jingfu Jia, Tao Cheng and Yaping Zhao, Polymeric coating of fluidizing nano-curcumin via anti-solvent supercritical method for sustained release, pp. 99-105, © 2014, with permission from Elsevier.

Zahibi et al. [174] have prepared curcumin/PLGA samples by a supercritical CO_2 anti-solvent procedure; the CO_2 flow was 2.5 ml min^{-1} at 33 °C and 9 MPa. A 1:4 ethanol:ethylacetate mixture was used as the solvent and the drug:polymer ratios tested were 1:2 and 1:1. Homogeneous distribution of the curcumin particles within the polymer was favoured by application of ultrasounds [175]. As the curcumin/PLGA ratio was increased, the size of the particles decreased and the amount of drug loaded was increased. Release from these systems was sustained, 13.8 % of the initial amount of drug was released within the first 200 min, and 27.7 % after 400 min, without any burst effect during the first release stage, Figure 7.

Layered Double Hydroxides As Drug Supports

Biologically-active molecules can be inserted in LDHs. This subject has been recently reviewed by different authors and thus only some highlights will be mentioned here. One of the first reviews on this subject was published in 2001 by Costantino and Nocchetti [176], and it was followed by several other from different research teams [9,10,59,127-129,177-184]. In these studies the subjects reviewed have been the intercalation of antibiotics, anticancer agents, vitamins, non-steroideal antiinflamatory agents, amino acids and peptides, anticardiovascular agents, antifibrinolytic agents, antihypertensives, liposomes, antimycotic agents, anticoagulants, antioxidants, inmunosuppresant corticosteroids, nucleosides and agents for diabetes and osteoporosis.

LDHs are used in cosmetics to remove inflammatory exudates found in acne vulgaris, to encapsulate organic molecules, or for protection against solar damage of the skin, as well as against degradation by light, heat or oxygen of labile molecules.

Among different antacids, MgAl-CO3 shows a larger neutralization capacity than conventional antacids and a prolonged buffer effect in an optimum pH range [126,185,186] with no dangerous increase of aluminium levels in serum [186]; these antacid properties are however dependent on the route to prepare the LDH. Hydrotalcite is present in different commercial drugs, such as Talcid© or Almax©, and MgAl-LDHs have been also used to adsorb intestinal phosphate to prevent hyperphosphatemia [187].

Because of their ease preparation, low cost, good biocompatibility, low cytotoxicity, and full protection of the loaded drug, hydrotalcite-type materials have been also proposed for the controlled release of intercalated drugs, what does not only reduce stomach irritation but increases the drug solubility [155,188-192], Figure 8. The ion exchange capability of LDHs allows encapsulation of functional negatively charged biomolecules. It is generally observed that release of a LDH-loaded drug follows a two steps process, first a rapid release, and secondly a maintained, slow release, usually related to anionic exchange with anions of the medium.

The use of these supports or matrices aims to improve the release of drugs whichever the clinical stage of the patient or disease, as their use will maintain high, constant levels of the drug because of an improved farmacokinetics behaviour; as a result, dosage will be more easily improved and the dosing intervals could be expanded.

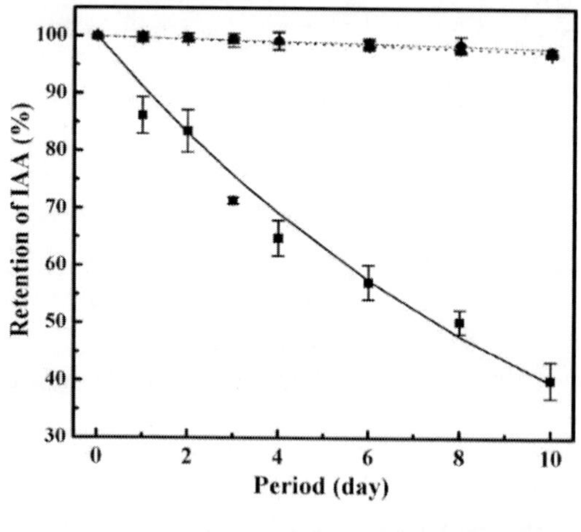

Figure 8. Retention of indole-3-acetic acid (IAA) for aqueous solution of IAA (squares) and the nanohybrid of IAA-Zinc based hydroxide salt (circles) and IAA-Zinc aluminum LDH (triangles) at 42 ºC in aqueous medium. Reprinted with permission from New inorganic-based drug delivery system of indole-3-acetic acid-layered metal hydroxide nanohybrids with controlled release rate, Jae-Hun Yang, Yang-Su Han, Man Park, Taeun Park, Seong-Ju Hwang, Jin-Ho Choy, Chemistry of Materials 2007, 19, 2679-2685. © (2007) American Chemical Society.

Among the different procedures applied to intercalate drugs into LDHs, larger loadings are reached when using coprecipitation (direct synthesis) or anionic exchange, but in this case a small amount of the anion existing in the original precursor usually remains, so it is recommended to use a chloride precursor to avoid the presence of toxic nitrate species. The reconstruction method also permits intercalation of drugs, but a biphasic system is usually obtained (LDH-carbonate and LDH-drug) with lower drug loadings because of incomplete removal of carbonate or its reintercalation upon CO_2 adsorption from the atmosphere. The strong electrostatic interactions between the drug and the layers lead to an increase in the thermal stability of the drug.

The drug loading depends on the chemical composition of the matrix, method and synthesis experimental conditions, charge layer density, etc., possibility to modify the orientation of the drug molecules in the interlayer, their release rate and the amount of released drug. When intercalated, the drug molecules usually form antiparallel bilayers, sometimes slightly tilted, with the carboxylate groups pointing towards the inorganic layers.

Release of the drug is fast at low pH because of dissolution of the matrix, while at pH values around 7 release takes place *via* anionic exchange with the anions existing in the liquid medium. The basic nature of the LDH might be also responsible for the larger solubility of the drug, even when they are simply mixed and not intercalated. The release rate depends on the particle size, crystalline morphology and aggregation state of the particles, which can be modified or even tuned during preparation. The release rate and the amount of drug released depend on the chemical composition of the matrix and on the synthesis variables mentioned above.

Despite these numerous advantages, these systems can hardly be used in oral administration, as they are very sensitive to the acidic medium, unless they are covered with a polymer film which preserves them from being acidically degraded. So, this is a field where new insights are still required; some studies are already available on the use of PLA or Eudragit© to cover and to protect the LDH-drug hybrid from degradation, as discussed in the following sections in this Chapter.

Another very interesting field where developments are expected are those related to the use of LDH-drug composites, together with a magnetic [193] nucleus for precise delivery in locations of the body where the action is required. This procedure will permit a faster action of the drug, as well as decreasing the total dose, thus avoiding undesired side-effects.

LDH-Supported Drugs Dispersed in PLA

Interest in layered filler/polymer nanocomposites (NCs) has developed very rapidly in the last 10-15 years, both in industry and academia. The trigger for this interest is the potential applications of these systems with better physical, thermal, and mechanical properties than conventional NCs [194]. This improved behaviour is directly related to the dimensions and microstructure of the dispersed phase [195]. The most convenient layered materials to prepare polymer NCs are those with layers possessing a width of 0.25 – 0.7 nm [196]. Materials suitable to prepare these NCs include smectites, syntetic clays, acidic layered silicates, layered hydroxides, layered double hydroxides, phosphates or phosphonates of tetravalent metal cations, halides, sulphides, cyanides, oxides, etc. [197]. In the specific case of LDHs, their delamination produces positively charged thin platelets with a thickness of a few atomic planes [198].

The structure and properties of a given NC mostly depend on the dispersion degree of the layered material nanoparticles within the polymer matrix. Two main types of morphologies can be identified: *intercalated* and *exfoliated* ones [194]. When the polymer and the filler are highly chemically incompatible, another type of morphology develops, closely related to a conventional composite (sometimes called *microcomposites* [199]), where the polymer simply surrounds the filler particles, but without really penetrating between the layers.

The *intercalated* structure is formed when the polymer penetrates between the layers of the clay leading to a highly ordered state, while the *exfoliated* structure corresponds to the case where isolated layers are dispersed in the polymer matrix [200]. In this structure the polymer/clay interactions are maximized, and all the surface of the filler particles might interact with the polymer; consequently, a drastic change in the physical, thermal and mechanical properties of the polymer is usually found.

Most of the studies concerning the use of layered inorganic fillers have been carried out with montmorillonite [43,201,202], an easily exfoliable layered silicate with low charge density layers. The main drawback is that montmorillonite being a natural compound, its chemical composition varies and the presence of impurities or different metal cations makes somewhat difficult to control and to reproduce the final properties of the NC [203]. For this and other reasons outlined below, an increasing use of layered double hydroxides is currently observed [56,197].

As mentioned above, LDHs display several advantages against conventional clays or layered silicates [204]:

(i) They are easily synthetized in the laboratory, highly pure and structurally homogeneous; their chemical composition can be tuned within a broad range depending on the pursued properties.

(ii) LDHs can be modified with anionic surfactants, and many different chemical compounds (e g., fatty acids, sulfonates, sulfates, carboxylates, phosphonates, etc.) result capable of being used, permitting to choose that most compatible with the polymer matrix.

(iii) The reinforcement ability of clay/polymer NCs depends on the exfoliation degree: A NC containing fully exfoliated LDH particles should contain a larger number of layers per mass of filler than a NC with the same amount (mass percentage) of silicate filler; as a result, the LDH reinforcement ability will be larger even for a lower filler loading [204]. This is because LDHs are formed by single layers of metal hydroxide octahedra, while the silicate layers are formed by double sheets of octahedra and tetrahedra.

(iv) Upon chemical decomposition, LDHs undergo endothermic processes releasing water and forming metal oxides, providing materials which are more efficient to improve the thermal stability and the resistance to fire [205].

Synthesis of LDH/Polymer Nanocomposites

Despite many methods have been proposed in the literature to prepare LDH/polymer NCs, most of the papers follow one of these [206]: *in situ* polymerization, melt compounding and solvent casting.

In Situ Polymerization

This method was first used to prepare a clay/polymer nanocomposite with polyamide-6 (PA6) [207]. Swelling of the layers was attained by intercalating an anionic surfactant, which also improved the LDH-polymer chemical compatibility [208]. Secondly, a suspension of these nanoparticles in a solution of the monomer (ultrasound-assisted in some cases) was prepared, and the polymerization reaction was initiated by heating, irradiation or any other way. As the chains of the polymer grew once the nanoparticles were within the polymerizing medium, a high dispersion degree is usually reached [209].

Melt Compounding

A milled polymer - LDH mixture was heated at a temperature slightly above the melting point of the polymer, and the *melted* polymer chains were inserted between the layers of the LDH [210]. Dispersion and exfoliation were not as effective as in the *in situ* polymerizartion method [206], and rather frequently the nanocomposites were only partially intercalated and exfoliated.

Solvent Casting

The use of suitable solvents permitted to swell and to disperse the LDH filler, which was even exfoliated in some cases [142]. The suspension was dispersed in a solution of the polymer and then this was precipitated by changing the medium. Due to the low price of water and its lack of polluting effects, the method can be used to prepare nanocomposites which polymer matrix is soluble in water. When other solvents are required, the cost of the process increases and environmental (disposal) and health problems arise.

Drug-LDH/PLA Systems for Drug Controlled Drug Delivery

As LDHs can be dispersed in PLA, and LDHs have shown their ability to host and to deliver drugs, studies on LDH-drug/PLA systems have been reported, aiming to combine the advantageous properties of the three-components systems.

Zhao et al. [211] have prepared MgAl LDHs with dodecyl sulphate (DS) in the interlayer by a conventional coprecipitation method. The spacing corresponding to diffraction by basal planes (003) was 25.4 Å, much larger than the value (7.8 Å) measured for a carbonate-containing sample, indicating that intercalation has been achieved and that the DS molecules were not oriented parallel to the brucite-like layers. A composite with PLA (MgAl-DS/PLA) was prepared by electro-spinning. Formation of these composites was confirmed by PXRD, which showed a broad maximum due to semicrystalline PLA and two weak maxima at 29.6 and 11.9 Å, ascribed to diffraction by planes (003) and (006) of the LDH phase, respectively, although the relationship between these two values did not match the expected value, probably to the broadness of the second maximum. The MgAl-DS/PLA system showed a rough surface with small knobs observable by SEM, suggesting that the hydrotalcite phase was not completely dispersed, forming conglomerated nanoparticles which have not been dispersed completely in the PLA phase. Thermal analysis showed that destruction of PLA took place at 145 or 150 ºC, depending on the shape (fibers or pellets, respectively) of the sample. The decomposition temperature decreased upon incorporation of increasing amounts of the LDH phase. However, these authors have also reported a change in the molecular mass of PLA and thus the change in the degradation temperature could be also related to a change in the orientation of the PLA crystalline fibers upon incorporation of hydrotalcite. In other words, degradation of PLA was favoured by the presence of the hydrotalcite phase.

Pan et al. [212] have also prepared a MgAl-DS/PLA composite by two steps. First, a nitrate sample (MgAl-NO3) was prepared by coprecipitation and then anionic exchange was used to remove the interlayer nitrate anions and to substitute them by DS. Different loadings (2.5, 5, 10, and 15 % w/w) of MgAl-DS hydrotalcite were incorporated into PLA by stirring at 80 rpm for 3 h at 190 ºC, just above the melting point of PLA. The resulting solid showed a spacing of 27.8 Å, a value 1.8 Å smaller than that reported by Zhao et al. [211] for similar samples, for diffraction by planes (003) of the hydrotalcite phase, while the spacing for the starting nitrate phase, MgAl-NO3, was 7.8 Å, suggesting that the nitrate anions were intercalated with their molecular plane parallel to the brucite-like layers, and confirming intercalation of the DS anions. The PXRD diagram of the LDH/PLA composite showed an increase in the intensity of a broad diffraction at 35.5 Å as the LDH loading was increased, while the spacing for planes (003) of the hydrotalcite had been recorded at 27.8 Å. Intercalation was further confirmed by the observed shift in the positions of the FTIR absorption bands of DS when compared to those of pure DS [213]. TEM micrographs showed additionally that some layers were exfoliated, coexisting with intercalated phases. The presence of the LDH favoured nucleation of PLA and decreased the size of the spherulites.

LDH/PLA systems have been used by Ha and Xhanthos as plastifiers [214]. A carbonate containing MgAl hydrotalcite was used as a precursor to prepare samples with intercalated phosphinates of different chain lengths by reconstruction [79,80] and the resulting product was mixed with PLA (5 % w/w) in a mixer at 50 rpm and 170 ºC for 7 min in a nitrogen atmosphere. TG studies showed a decrease in the thermal stability of the nanocomposite: while T_{50} (i. e., the temperature required to loose 50 % of the initial mass of sample) was 364 ºC for pristine PLA, it was only 299 ºC for the nanocomposite. The plastifier and degrading properties of the polymer were modified too by the presence of the LDH and the melt flow index values were rather large for the nanocomposite (low viscosity), decreasing the easiness for releasing the interlayer anion; in other words, incorporation of the nanocomposite allowed a fine control of phosphinate species release from the interlayer of the hydrotalcite.

Chiang and Wu [215] have prepared a MgAl-NO3 LDH by coprecipitation and then have applied an anionic exchange process to incorporate PLA-COOH, resulting in a solid wih a basal spacing of 14.03 Å (suggesting formation of a bilayer or a tilted layer of PLA-COO$^-$ units), which increased up to 48 Å when a well crystallized nanocomposite with PLA was formed; this increase confirmed intercalation of PLA in the interlayer space. In agreement with Ha and Xhanthos [214], the thermal stability of PLA decreased upon incorporation of the LDH, the temperature at which degradation started decreasing 80 ºC with respect to pure PLA; a similar thermal stability decrease was observed when the nanofiller loading was increased, Figure 9. It seemed [216] that the metal cations from the hydrotalcite layers catalysed the thermal degradation of the polymer by favouring depolymerization and/or inter and intra-transesterification processes, mainly leading to formation of lactones [38,150].

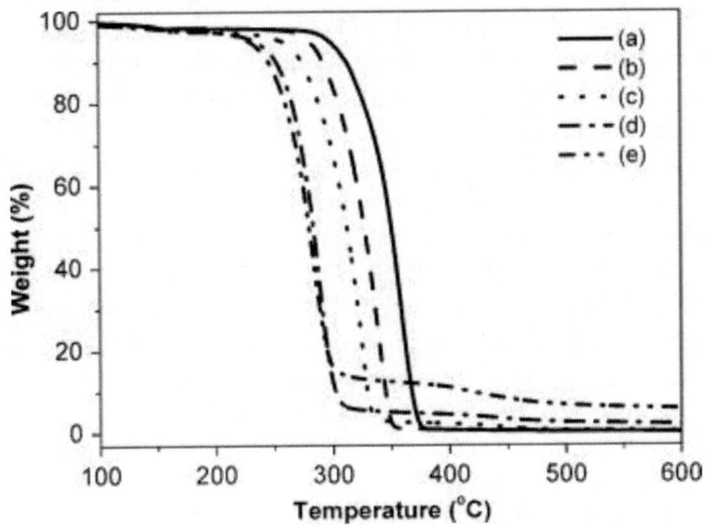

Figure 9. Thermogravimetric analysis patterns of: (a) neat PLLA matrix, (b) 1, (c) 3, (d) 5 and (e) 10 wt.% PLLA/PLA-COOH modified LDH. Reprinted from Composites Science and Technology, Vol. 70, Ming-Feng Chiang and Tzong-Ming Wu, Synthesis and characterization of biodegradable poly(L-lactide)/layered double hydroxide nanocomposites, pp. 110-115. © 2010, with permission from Elsevier.

Katiyar et al. [217] have prepared systems with intercalated PLA. They first prepared a MgAl-laurate (C12) LDH by reconstruction from a MgAl (3:1) hydrotalcite. Then PLA was intercalated by *in situ* polymerization of lactic acid catalysed by tin(II) 2-ethyl hexanoate. Different LDH loadings had only a minor effect on its own swelling, the basal spacing increasing from 7.8 Å for the starting MgAl-CO3 precursor to 25.6 Å for the laurate derivative. Intercalation of PLA led to complete exfoliation of the nanofiller, as confirmed by TEM, and no diffraction maximum attributable to diffraction by basal planes of the LDH was recorded. The MgAl-laurate/PLA systems showed a high crystallinity of the PLA matrix, suggesting a positive effect of the hydrotalcite on the crystallization process. This *in situ* polymerization led to PLA with a lower molecular mass than in the absence of the LDH; it seems that the hydroxyl groups of the hydrotalcite crystallites promoted termination of the polymer chain, while the metal cations favoured degradation of PLA.

Wang et al. [218] prepared a ZnAl-DBS LDH with a basal spacing of 30.3 Å; when mixed with PLA only diffractions due to PLA were observed, but no clear diffraction maxima due to the LDH was observed, probably due to exfoliation of the layers and formation of tiny tactoids, as evidenced by TEM. The flame retardant effect of the LDH in this nanohybrid has been checked also by these authors.

Gerds et al. [219] have claimed that degradation of PLA in MgAl-laurate/PLA systems depended on the preparation procedure. Even when using samples with very low loadings of the inorganic filler (1 %), degradation was easier for samples prepared by reconstruction (using sodium laurate) than for those prepared by coprecipitation, using lauric acid in a NaOH(aq) medium. Probably the lower crystallinity of the samples prepared by reconstruction made them more active to favour degradation of the polymer. The good dispersion and miscibility of the hydrotalcite in the PLA matrix promoted the interaction with the hydroxyl groups, favouring degradation of the PLA [220]. The PLA-LDH interface area increased as the dispersion of the LDH did, and thus depolymerization was favoured by the hydroxyl groups bonded to the Mg^{II} cations. LDH dispersion in the PLA matrix, which influenced the accessibility of PLA to the hydroxyl layer surfaces and to catalytically active Mg^{II} sites, seemed to be in the origin of PLA degradation.

ZnAl LDHs (Zn:Al molar ratio 3:1) with intercalated stearate (C18) have been prepared by Eili et al. [221] by coprecipitation and by anionic exchange using a LDH precursor in the nitrate form. Intercalation of stearate led to swelling of the LDH layers, reaching a basal spacing of 40.1 Å, as evidenced by PXRD experiments. The LDH systems were mixed with PLA in the presence of chloroform and nanocomposite films were obtained upon evaporation of the solvent. When different LDH loadings were used, even for 10 % (w/w) levels, the signals recorded in the PXRD diagram corresponded exclusively to PLA. However, and as mentioned above for other systems, degradation of PLA (as studied by TG-DTA) started at a lower temperature than for pristine PLA when LDH was added; this effect was attributed to the presence of the long hydrocarbonated chains of the stearate anions, which could favour mobility of the PLA chains, decreasing their strength and thoughness. At room temperature, the degradation rate of PLA increased with the LDH loading; degradation of the stearate units favoured a decrease in the molecular mass of the polymer, PLA chains breaking in small particles, and thus increasing the overall degradation.

Dagnon et al. [152] have reported studies on systems prepared with ZnAl LDHs containing IBU in the interlayer space. A ZnAl-NO3 precursor was first prepared, and IBU was incorporated by anionic exchange. The resulting solid was mixed with PLA (loadings of the LDH were 1, 3, or 5 % w/w) and the mixture vigorously stirred at 40 ºC for 4 h. Films of the nanocomposite were prepared with a custom-built spin caster without needing any substrate. The interlayer spacing reached a value of 22.95 Å, confirming intercalation of the IBU molecules. When PLA fibers were used the spacing further increased to values of 27.82 and 27.99 Å for samples with loadings of 1 and 3 %, respectively. The fibers were also apparently appart, according to TEM micrographs, upon incorporation of the LDH, but the dispersion resulted less homogeneous when the concentration of the inorganic component was incresased. Incorportion of the LDH had a negative effect on the thermal stability of PLA; the DTG diagram for pure PLA showed two mass losses at 120 ºC and 387 ºC, but the second signal decreased even to 331 ºC for the sample with 5 % (w/w) of LDH, probably because water evolved during decomposition of the LDH favoured the hydrolitic decomposition of PLA. Alternatively, the polymer matrix may become unstabilized by the presence of the inorganic nanofiller, thus favouring its thermal decomposition at a lower temperature. These authors have studied also the release of the drug; it followed a well know pattern, with a fast release during the first 15 h, followed by a slow release; also the percentage of drug released increased with the drug loading, but even for the sample with 5 % LDH loading, only 70 % of the initial amount of drug was released after 50 h.

MgAl LDHs with IBU have been also studied by Miao et al. [153]; the LDH samples were prepared by coprecipitation, and mixed with PLA by electrospinning. The basal spacing for MgAl-IBU was 21.0 Å, rather similar to the value reported by Dagnon et al. [152], 22.95 Å. PLA was responsible only for a very broad diffraction at $2\theta=15°$, suggesting a low crystallinity of the PLA matrix. However, if the nanocomposite was formed with polycaprolactone (PCL) or PLA/pluronic (triblock copolymer based on poly(ethylene glycol)-poly(propylene glycol)-poly(ethylene glycol)) instead of PLA, the basal spacing increased up to 28.8 Å, probably because PCL was partially intercalated within the LDH layers. Thermal degradation of the polymer (PLA or PCL) was not modified upon incorporation of non intercalated IBU, probably because of its low loading (2% IBU/PLA or 5% LDH-IBU/PLA). Degradation of the LDH/PLA (or LDH-PCL) nanohybrids started ca. 80 ºC below the temperature at which it was observed for the pure polymers.

Release of IBU from PLA was slower than from the nanohybrid with PLA or PLA/pluronic mixtures. This difference could arise from the fact that PLA is strongly hydrophobic; degradating interaction of water with the polymer was hindered because of the low flexibility of the PLA chains, making difficult the mobility and difusion of water molecules. The presence of the hydrotalcite phase within the PLA matrix expanded the PLA fibers, favouring water difusion and thus favouring release of IBU. Incorporation of 25 % pluronic increased the hydrophylic character of the system and the release rate of IBU increased by 80 %. This behaviour had been previously observed by other authors with PLA-containing systems [222,223].

A ZnAl-LDH with intercalated IBU was prepared by anion exchange from a nitrate precursor by DeLeon et al. [154]; then it was mixed, 3 and 5 % (w/w), with PLA by spin casting and the mixture was dried for 2 days at 35 ºC. The ZnAl-IBU sample showed a basal spacing of 24.7 Å, somewhat larger than the values reported by other authors (see above), suggesting formation of an alternate bilayer in the interlayer space as confirmed by FTIR spectroscopy. In the nanohybrids with PLA the basal spacing slightly increased to 26.6 Å (for the sample containing 5 % LDH).

Ibuprofen release was monitored by HPLC. Release was rather fast during the first 15 h (difusion-controlled), but slowly from 50 to 144 h, due to anionic exchange [224]. Release from the LDH-PLA system was controlled by degradation of the polymer [153]; however, and also in agreement with other results above commented, the release rate was slower when the drug was directly mixed with PLA without being intercalated in the LDH phase, confirming the role of the inorganic component on the degradation of the PLA matrix and the drug release rate.

San Román et al. [225,226] have prepared ZnAl LDHs with chloramphenicol (CHLO), diclofenac (DICLO) and ketoprofen (KET) in the interlayer space by coprecipitation. The basal spacings were, respectively, 11.65, 23.35, and 21.97 Å, suggesting formation of a slightly tilted bilayer in the two last cases. On mixing the LDH with PLA (4 % w/w, heating at 80 ºC for 2 h) diffraction maxima due to semicrystalline PLA were also recorded, and the maxima due to the LDHs did not shift; only in the case of the ketoprofen-containing sample a weak maximum at 24 Å was additionaly recorded. This finding was similar to those previously reported by Ha and Xanthos [227] and Bugatti et al. [228] and has been ascribed to insertion of PLA in the interlayer space of the LDH. Thermal studies on these samples [226] reported again that the polymer was degraded at a lower temperature than in the pristine state, Figure 10. Release of the drugs in a PBS solution at 37 ºC from the LDHs was fast and completed after 24 h. However, for the LDH/PLA systems the release was markedly slower and could be modulated: after 12 weeks, only 70 % of CHLO was released, and the released percentages of KET and DICLO were even lower, 36 and 24 %, respectively. These results indicated that the amount of drug released depended on the precise nature of the intercalated anion and on the easiness of its diffusion through the polymer [229].

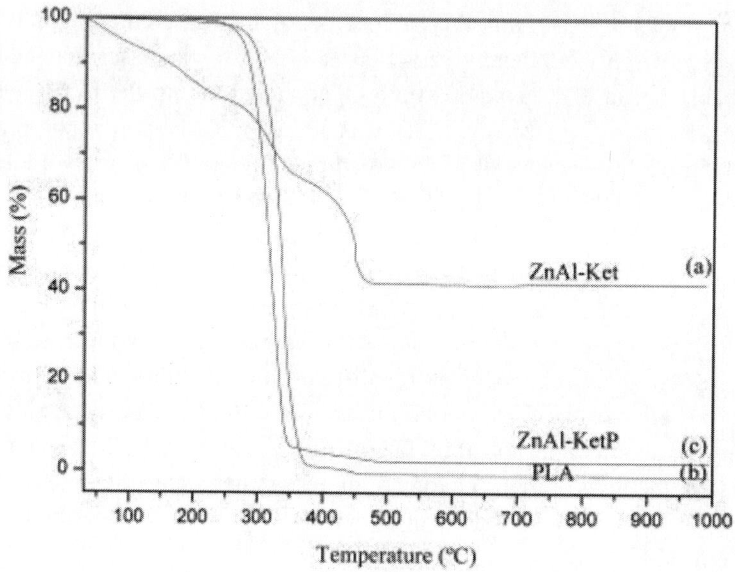

Figure 10. Thermogravimetric analysis curves of ketoprofen intercalated in a ZnAl LDH (ZnAl-Ket), of polylactic acid (PLA), and of the ketoprofen-LDH/PLA nanocomposite. Reprinted from Applied Clay Science, Vol. 71, María S. San Román, María J. Holgado, Beatriz Salinas and Vicente Rives, Drug release from layered double hydroxides and from their polylactic acid (PLA) nanocomposites, pp. 1-7, © 2013, with permission from Elsevier.

LDH-Supported Drugs Dispersed in Other Polymers

The first papers published on these systems were reported by Costantino et al. [230] and gave rise to a large amount of research in this field.

These authors [230] prepared a LDH with MgAl and 12-hydroxydodecanoate (HD) in the interlayer by ion exchange from a nitrate-containing LDH, and then prepared several nanocomposites with poly (ε-caprolactone), PCL, by (a) solvent casting, (b) ring-opening polymerization (ROP) of ε-caprolactone, and (c) addition of a high molecular mass (80 kDa) PCL to the sample prepared by procedure (b). Route (a) led to solids which PXRD diagram showed the unshifted maxima of the original LDH phase; the diagram for samples prepared by route (b) showed a sharp maximum at 14 Å (it was at 22.7 Å for the original MgAl-HD LDH), which probably is due to intercalation of PCL oligomers, or a partial HD/6-hydroxycaproate exchange, this last species formed *via* hydrolysis of ε-caprolactone [231]. Finally, no diffraction maximum was recorded at low diffraction angles for the sample prepared following route (c), suggesting complete exfoliation of the LDH as a result of the interactions between the intercalated PCL chains and the chains of the added high molecular mass polymer. These results were confirmed by FTIR spectroscopy.

Costantino et al. [232] have also prepared a MgAl-NO3 hydrotalcite, where interlayer nitrate was replaced by CHLO by ion exchange. Different amounts of the LDH-drug compound were incorporated into polymeric ε-caprolactone and it was found by TG that thermal degradation of the nanocomposites, whichever the amount of LDH added, started ca.

44 ºC before than for pure ε-caprolactone. The PXRD diagram suggested that the long range order of the LDH decreased when mixed in the nanohybrid, as the diffraction lines were broader, and even in some cases no maxima due to basal planes was detected, suggesting a complete exfoliation of the LDH.

Release of CHLO was monitored by UV-Vis spectroscopy. After a first, fast, release step, a second slow process was observed; in both cases, the release rate depended on the amount of LDH initially existing in the nanocomposite; similar results were reported for intercalated DICLO [233]. The authors suggested that the first stage corresponded to exchange of CHLO molecules anchored to the inorganic layers on the surface of the samples, while the CHLO molecules anchored inside the film were released in the second release stage. In addition to the exchange process, release could be also accomplished by diffusion of anions and counter-cations which ought to be exchanged. This interpretation is supported by the fact that the fraction of drug released in the firt stage depended on the sample thickness. This release mechanism was different and much slower than in those samples where the antibiotic molecule was directly inserted in the polymer matrix, without the inorganic component.

Similar systems containing a MgAl-DICLO LDH in a polycaprolactone matrix have been reported by Costantino et al. [229]. First of all, these authors found that the mechanical properties, except fracture, were improved upon incorporation of the inorganic component. Concerning DICLO release in a physiological medium (0.9 % w/w NaCl in water) from samples prepared as films, it took place in two consecutive steps, first a fast release, very likely from the external surface of the lamellae, followed by a second, much slower process ascribable to deintercalation of the drug inside the polymeric film.

Costantino et al. [234] prepared ZnAl (2:1) LDHs following the urea method [86] and tranexamate anions were intercalated by ion exchange. The solid thus obtained was added in different amounts into poly (ε-caprolactone) by mixing and milling in a ball mill at 580 rpm for 1 h; for a reference, the unloaded polymer was also submitted to this mechanical treatment. Films were prepared in a press between two layers of teflon. PXRD showed that the polymeric chains were partially intercalated in the LDH particles [230], although this effect was not observed in other cases [232].

As in previous studies [232,235], TG analyses showed [234] that thermal decomposition of the polymer in the nanohybrids started at lower temperatures than for the pure polymer, probably because a base-catalysed hydrolysis of the ester bonds by water, released from the LDH during the first steps of the heating treatment. Quite originally, the release of the drug from these nanohybrids was followed by FTIR spectroscopy (both in the solid and in the physiological solution) by measuring the intensity of the absorption band close to 1535 cm^{-1} due to the carboxylate moiety of the drug molecule, and not by UV-Vis spectroscopy monitoring of the drug in the solution. As expected, two release stages were observed: a first, very fast, almost burst, stage, followed by a second much slower one showing a *plateau* from 100 days onwards, but even then only 40 % of the original drug loading was released.

Antimicrobial agents, i. e., benzoate (Bz), 2,4-dichlorobenzoate (BzDC), and p- and o-hydroxybenzoate (p- and o-BzOH) have been intercalated in ZnAl (2:1) LDHs by anionic exchange from a nitrate precursor. These were then incorporated into poly(ε-caprolactone) [236], Figure 11, by mixing PCL with different amounts of the LDH and submitting the mixture to high energy ball milling (HEBM) [237]; the solids prepared were processed to films ca. 100 μm width. The PXRD patterns of the nanohybrids with Bz, BzDc and o-BzOH showed a small shift of the maxima due to diffraction by basal planes to lower diffraction angles, suggesting a partial intercalation of the PCL chains within the interlayer space of the LDH. The heigth of the first (at low diffraction angles) diffraction maximum increased as the LDH loading was increased. However the maxima are weaker and broader than for the original LDH, well because of partial exfoliation, or crystallite size reduction by the milling process.

However, two phases, the LDH and the pure polymer, were identified for the sample containing p-hydroxybenzoate, suggesting the lack of exfoliation or intercalation in this sample. The authors assumed that the lateral π-π interactions between the benzene rings of the intercalated anions were rather weak, and the PCL chains could somewhat destroy the layered structure; however, for the sample containing o-BzOH the hydroxyl groups could stablish stronger hydrogen bonds with the layers.

According to TEM results, the films were opaque, homogeneous, white and without clusters, while a set of samples prepared by physical mixture for comparison were non homogeneous and some clusters could be observed. The thermal decomposition was also anticipated to the temperature at which it took place for the pure polymer, and this finding has been ascribed also by these authors to hydrolysis by water released during decomposition of the LDH. Bugatti et al. [228] studied the release of benzoate from these samples; it was slower when the LDH/PCL was used than for a benzoate-PCL mixture. This finding is very interesting for using these systems in the active packaging field, where the antimicrobial drug should be retained by the polymer for a long period of time. The activity of the released antimicrobial agents was essayed by using *Saccharomyces cerevisiae* microorganisms; the growth of the culture was inhibited by 40 % on comparison with an untreated culture.

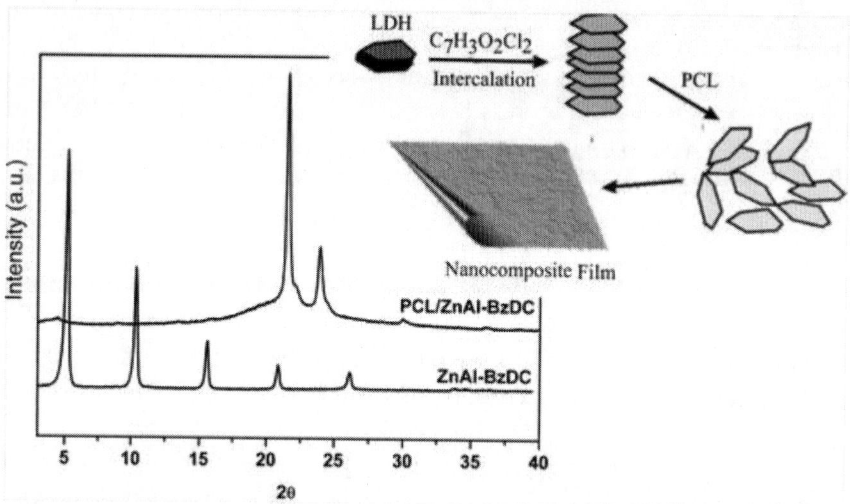

Figure 11. Scheme depicting formation of polymeric composites based on poly(ε-caprolactone), PCL, and layered double hydroxides containing benzoate-derived antimicrobial species, and powder X-ray diffraction profiles of a dichlorobenzoate (BzDC)-intercalated ZnAl LDH (bottom) and of its composite with PCL. Reprinted with permission from New polymeric composites based on poly(ε-caprolactone) and layered double hydroxides containing antimicrobial species, Umberto Costantino,Valeria Bugatti, Giulana Gorrasi, Francesca Montanari, Morena Nocchetti, Loredana Tammaro, Vittoria Vittoria, Applied Materials & Interfaces 2009, 1, 668-677. © (2009) American Chemical Society.

Concerning the samples with 2,4-dichlorobenzoate and p-hydroxybenzoate, the release was studied [238] from three different samples, namely, from that with the antimicrobial agents directly dispersed in the polymer (PCL), from the exfoliated LDH dispersed in the polymer, and from LDH microcomposites in the polymer matrix. Release was fast and in a single step in the first case, but much slower and in two stages (a first fast one, lasting for several days, followed by a slow one, lasting for several months) from the nanohybrids; in this last case, total release was never achieved. These authors have also claimed that in the absence of anions in the release medium, the process did not take place (suggesting release was due to anionic exchange), and that in the p-hydroxybenzoate containing sample the strong hydrogen bond interactions between the hydroxyl groups of the anion and the LDH layers retained the drug rather strongly, thus accounting for the lack of total release [239].

The mechanical properties of these samples were studied by Bugatti et al. [228]. From the experimental stress-strain curves these authors concluded an improvement or, at least, a lack of declining of the mechanical parameters; the barrier properties were measured for the sorption isotherms of water vapour; all composites showed a behaviour similar to that of pure PCL, but with a slightly larger absorption ability, probably due to the hydrophilicity of the hydrotalcite layers, finding a significant relationship between the water absorption capacity and the LDH content in the nanocomposite.

Miao et al. [153] have prepared MgAl LDHs with intercalated IBU by coprecipitation and then have prepared nanocomposites with PCL by electrospinning, following a procedure similar to that above mentioned using PLA as a polymer matrix. TEM showed a homogeneous dispersion of the LDH in the polymer, contrary to other findings with MgAl-12-hydroxydodecanoate or MgAl-dodecylsulphate [211,240], probably because of the different size of the fibers and to the role of IBU, acting as an organic surfactant, which improved the chemical compatibility between the LDH particles and the polymer matrix. The PXRD diagrams showed that direct dispersion of IBU in the polymer was homogeneous, with no diffraction maximum due to the dispersed drug; on dispersing the ibuprofen-containing LDH in the polymer a weak maximum at 28.8 Å was recorded (the value was 21 Å for the undispersed LDH), suggesting additional intercalation of PCL chains or even incipient exfoliation of the layers, contrary to the findings observed when the polymer was PLA, see above [153].

As in the studies of Bugatti et al. above commented, release of the drug from the IBU/PCL system was faster than from the LDH/PCL system, 40 and 75 %, after 2 h, respectively, such differences being also observed after 12 h.

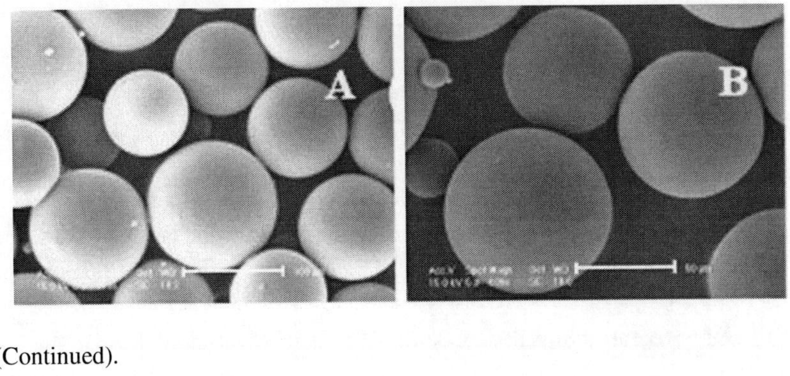

Figure 12. (Continued).

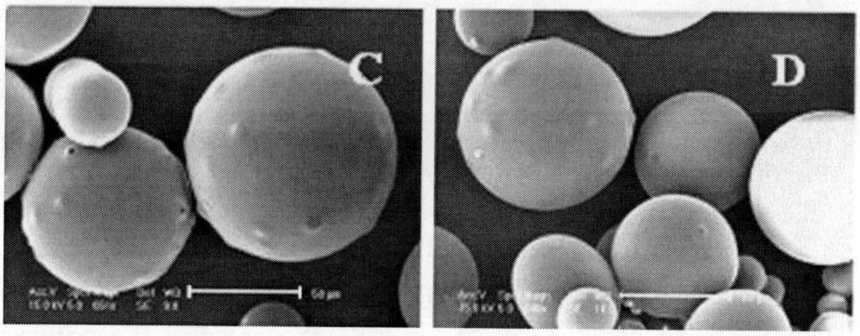

Figure 12. Scanning electron micrographs of diclofenac-intercalated LDHs dispersed in different Eudragit® polymers in different ratios: (A) Eudragit S 1:10; (B) Eudragit L 1:10, (C) Eudragit L 1:5, and (D) Eudragit S 1:5 microparticles. Reprinted from Microporous and Mesoporous Materials, Vol. 115, V. Ambrogi, L. Perioli, M. Ricci, L. Pulcini, M. Nocchetti, S. Giovagnoli and C. Rossi, Eudragit® and hydrotalcite-like anionic clay composite system for diclofenac colonic delivery, pp. 405-415. © 2008, with permission from Elsevier.

ZnAl LDHs containing amoxycillin (AMOX) have been prepared by coprecipitation [241] and encapsulated (in different concentrations) in PCL by electrospinning. A weak diffraction maximum at 23.2 Å was recorded, suggesting a loss of the basal periodicity of the LDH and that its layers were well dispersed in the PCL matrix. The thermal behaviour was similar to those shown by other LDHs dispersed in PCL, i.e., degradation took place at a lower temperature than for pristine PCL, and the decrease of the temperature at which the degradation started was larger as the LDH content was increased. SEM images showed fibrilar structures randomly oriented with a diameter close to 0.8 μm for pure PCL, increasing to 0.8 – 1.2 μm upon dispersion of the LDH-AMOX system. Release studies showed a first burst step followed by a slow second stage. The first step has been ascribed to release from the surface of the fibers and might be benefitial to eliminate the intruding bacteriae before they begin to proliferate, while the slow step would control the growing of the bacteria not eliminated in the first step.

Ambrogi et al. [242] have prepared and characterized a MgAl-diclofenac LDH supported on Eudragit® S or Eudragit® L using a modified oil-in-oil solvent evaporation method [243], with a 1:5 or 1:10 (w/w) LDH/polymer ratio. The SEM images showed the absence of aggregates; the particles were nicely spherical and with a rather homogeneous radius, Figure 12; when a larger amount of Eudragit® was used the surface was smoother, whichever form (S or L) of the polymer was used. The particle size distribution was dependent on the LDH:polymer, changing from 20-120 to 50-150 μm for 1:5 and 1:10 ratios, respectively.

In vitro release of DICLO was studied. Gastrointestinal conditions (37 ºC, at three pH values, 1.2, 6.8, and 7.5) were simulated, as the solubility of these polymers depends on the pH of the medium: both are insoluble at pH below 5, the L form is soluble at pH above 6 and the S form is soluble at pH larger than 7. As expected, no release was observed in an acid medium, but at pH=6.8 Eudragit® L was more efficient than Eudragit ® S, i. e., more drug was released, but at pH 7.5 the opposite behaviour was observed. Probably, a combined use of both forms of the polymer would permit selective release of drugs in different parts of the intestine.

In order to improve the solubility of some drugs in biological fluids, they have been intercalated in hydrotalcites. Choi et al. [244] have prepared a MgAl LDH by coprecipitation with intercalated ursodesoxicolate (UDC), currently used to dissolve cholesterol, and dispersed the compound formed in Eudragit ® S100. All experimental data (e.g., PXRD, FTIR spectroscopy) confirmed that the drug has been effectively intercalated, the basal spacing reaching a value of 33.12 Å, which was not modified upon dispersion in the polymer. In agreement with previous studies [245,246], no evidence was found of lixiviation of the drug upon incorporation of the LDH in Eudragit® S100. The release studies showed no significant release of the drug at pH=1.2, due to the lack of solubility of the polymer in this medium. Significant dissolution was observed from pH 6.8 upwards, reaching a value of 92 % of the initial amount of drug intercalated at pH 8 after 2 h. The Eudragit molecules could probably replace the intercalated drug molecules, swelling the interlayer space and thus favouring diffusion of the drug molecules to the liquid medium, although no PXRD data to confirm this assumption was provided. Parallel studies also in a physiological medium, but with samples without Eudragit, did not show release of the drug, as the lower size of the chloride anions would not swell the layers and diffusion of the drug molecules would be hindered [247].

Ha and Xanthos have prepared [227] MgAl LDHs with DICLO intercalated by reconstruction, and these systems have been supported on different Eudragit ® polymers, E100, S100 and S100 with added triethyl citrate (TEC). Samples containing diclofenac directly dispersed in the polymers and samples with different amounts of the LDH dispersed in the polymer were prepared. The composites were prepared by compounding in a Brabender batch mixer at 50 rpm and 130 ºC for 5 min; after melt mixing the compounds were pressed for 50 s at 130 ºC into thin disks. The PXRD diffractogram of the sample prepared with Eudragit® E100 (amorphous) showed a maximum due to diffraction by planes (003) of the LDH at 22.64 Å, very close to the value for the undispersed LDH (22.41 Å), suggesting that the polymer has partially accessed into the interlayer space, although the change is minimal. This was similar to the effect previously observed by these authors with samples supported on polypropylene [248].

Release of the drug studies were carried out in a solution simulating the gastric fluid at pH 1.2. Despite DICLO is not soluble, its solubility increased when it was supported (for instance, intercalated in the LDH phase), but not when merely mixed with the LDH. The behaviour observed for the samples containing Eudragit® was similar in all cases for the first release step, suggesting that the effect of degradation of Eudragit® is more important that retention by the LDH structure, due to the small size of the powder particles, but in no case total release of the drug was observed even at the end of the study (300 min). The samples prepared with Eudragit ® S100 showed a slower release of DICLO in the absence of LDH at pH 7.4. This fact can be related to the presence of "packets" of hydrotalcite nanoparticles within the polymer, modifying the release mechanism.

Rives et al. [249] have intercalated fenbufen (FENB) in MgAl and MgAlFe LDHs by different methods: coprecipitation, ion exchange from precursors in the chloride form, and reconstruction from the calcined carbonate form. Some of the samples were covered with Eudragit® S100 (insoluble at pH above 7) by two alternate methods: simple dispersion of FENB in the polymer matrix, or microspheres containing the free drug or the LDH-drug (prepared by ion exchange); the relase of the drug was also studied.

When FENB was dispersed in the microspheres, only the diffraction maxima of Eudragit® S100 were recorded in the PXRD diagram, but when FENB-LDH was dispersed, a diffraction maximum correspondig to planes (003) of the LDH-drug phase was recorded at low angles, indicating that formation of the microspheres did not destroy the layered structure [250]. According to SEM images, the surface of the microspheres was smooth and homogeneous, with an average size of 50-60 μm, suitable for oral administration; the sample prepared by simple dispersion in the polymer showed a very different aspect.

Release studies were carried out at pH 1.2 and 6.8 for 2 h and at pH 7.5 for 5 h, to simulate the conditions in the gastrointestinal tract. No release was observed at pH 1.2.; 54 % release was observed at pH 6.8 from the sample prepared by dispersion, and only at pH 7.5 a complete release was achieved from all samples, but faster from the microspheres, probably by ion exchange with phosphate anions from the buffer solution used. These results were different from those reported by Li et al. [251], but in that case microspheres were not formed, suggesting that the interaction of the carboxylate groups of the polymer with the surface of the hydrotalcite microcrystals inhibits the deintercalation of the drug and dissolution of the polymer.

NiMgAl LDHs with stearate were prepared by coprecipitation coupled to microwave hydrothermal treatment [252]; the solid was then supported on ethylene vinyl acetate (EVA). These authors studied the effect of the presence of nickel on the flame retardant and thermal properties of the composites prepared. The maximum mass loss rate temperature and the total mass loss were reduced as the nickel content was increased. The sample without Mg and a Ni/Al molar ratio of 3/1 showed the best results for flame retardance and suppression of smoke and toxic gases. Concerning the mechanical properties, the ductility and strength was much more improved when Ni was present than when a MgAl LDH was used.

Xu et al. [253] have prepared LDHs with Mg and Al and carbonate by urea hydrolysis [86], and then exchanged at pH 4 in a nitrogen atmosphere with nitrate, dodecyl sulphate (DS) and dodecyl benzene sulphonate (DBS); then the solids were dispersed in poly(ethylene terephthalate), PET. The typical broad "halo" of amorphous PET was recorded in the PXRD diagram of the composites; however, no other diffraction line was observed when the interlayer anions were the organic ones, suggesting exfoliation of the filler, but two diffraction lines at 9.01 and 4.5 Å due to the LDH-NO3 phase were recorded. These results were in agreement with the TEM images, which showed elongated particles of the nitrate-containing LDH, while for the organic derivatives only individual layers and small particles (ca. 100 nm) were observed. Contrary to other results with PCL or PLA as the polymeric component [153,232], the TG curves of pure PET and of the nanohybrids were almost coincident.

Liu et al. [254] have prepared LDHs with naphtalene acetate (NAA) from MgAl-NO3 LDHs by anion exchange; this compound has been used to prepare films by the layer-by-layer (LBL) self-assembling method, proposed by Decher [255] with cationic polyacrylamide, polyacrylic acid sodium and LDH. PXRD results showed an increase of the interlayer distance, with spacing for (003) increasing from 8.9 Å for the nitrate precursor to 20.9 Å for the LDH-NAA compound. As the thickness of the NAA molecule is 8.5 Å and of a brucite-like layer 4.8 Å [256], formation of a bilayer was proposed, where the carboxylate groups pointed towards the top and bottom hydroxyl layers. FTIR spectra confirmed intercalation of the drug, and the bands due to the symmetric and antisymmetric vibrations of the carboxylate groups and other bands due to the aromatic rings were recorded; however, the values for the film where somewhat different, probably due to the presence of the polymer. The morphology of the samples, according to SEM, corresponded to randomly oriented LDH particles with a non-regular shape and rough surface. The surface of the film was not uniform, probably due to non homogeneous sedimentation of the LDH particles.

The drug release studies were carried out at different temperatures and pH values (adjusted by addition of aqueous solutions of HNO_3 or NaOH), and the amount of NAA released was measured from the absorption at 282 nm in the UV-Vis spectrum. Release was slow and the amount released increased with the time the nanocomposite was in contact with the solution; an increase in the temperature also led to an increase in the amount of drug released, as it decreased the viscosity of the polymer and accelerated LDH difusion, inhibiting the initial burst step. Release was faster at pH 2 than at pH 9, as at the former pH an anion exchange of the intercalated organic anions and nitrate anions from the medium took place. When the release studies were carried out from the LDH without polymer, similar results were obtained at acidic or basic pH values, release taking place in the former case by destruction of the layered structure and at basic pH by anion exchange. Under all experimental conditions, release from the LDH/polymer films was slower than from the LDH, following a first order kinetics.

Ruiz-Hitzky et al. [257] prepared MgAl-LDHs with 5-amminosalicylic acid (5ASA) by coprecipitation, in order to be used for colon cancer treatment. The LDH was dispersed in chitosan as a polymeric matrix by slowly pouring the mixture in a 2M NaOH solution. The pearls thus formed were dispersed in the polysaccharide pectin to improve the stability of chitosan and LDH under acidic conditions, as found in gastric fluids, Figure 13. For comparative purposes, different samples were prepared: pure 5ASA, 5ASA/chitosan, on one side, and also pectin-encapsulated 5ASA, LDH-5ASA, 5ASA/chitosan and LDH-5ASA/chitosan. The samples containing pure, unintercalated drug were more efficient in encapsulation than those containing the LDH-5ASA hybrid, probably because a protective effect of the LDH layers against exit of the drug molecules during handling and purification of the pearls. A larger drug loading was observed for LDH/pectin than for LDH/chitosan, probably due to the alkaline medium of the starting pectin solution, thus hindering release of the drug from the LDH interlayer space.

The PXRD diagram of LDH-5ASA showed a basal diffraction at 20.8 Å, confirming intercalation of the drug; this means a distance between the layers of 16 Å (the brucite-like layers have a width of 4.8 Å) and, taking into account the molecular size of the drug, these authors concluded formation of a bilayer, in agreement with previous results by Zhang [258]. FTIR spectroscopy results also confirmed intercalation of the drug; the bands due to the primary ammine and the ammino group of chitosan are shifted towards lower wavenumbers, probably because a partial deprotonation of chitosan during formation of the pearl; however, the characteristic bands of the LDH remained unaltered, confirming the stability of the layered structure. The bands of pectin in the pectin-covered pearls were also shifted with respect to the spectrum of the calcium form of pectin, suggesting an interaction between the free carboxylate groups of pectin with the unprotonated ammino groups of chitosan, as already observed in polyelectrolyte systems [259]. SEM images confirmed sharp differences between the pearls, covered or uncovered with pectin: in the former case the particles were homogeneous, slightly rough and porous, while the uncovered particles showed a larger roughness.

Absorption of water is one of the crucial factor governing release of the drug, and it was studied by these authors at different pHs. The pearls with chitosan were completely dissolved at acidic pH after 30 min, but when covered with pectin total dissolution needed 6 h, confirming the protecting and stabilizing role of the pectin cover. The chitosan spheres were more resistant to swelling at pH 6.8 than the covered ones. Finally, the protecting effect of pectin was confirmed by following the *in vivo* release at pHs simulating the path of drugs through a human body. Total release of the drug from chitosan was observed at acidic pH, suggesting that the pearls would be decomposed in the stomach; no release was observed if protected with pectin and at basic pH the slow release of 5ASA was due to chitosan.

The Use of PLA Copolymers

In order to improve the ability of PLA for drug release, several systems consisting of D and L lactic and glycolic acid copolymers (in different molar ratios) have been investigated, as degradation of these copolymers is easier than that of the corresponding homopolymers [260]. The first copolymer with clinic uses was developed by Ethicon (VICRYL®) for quirurgical sutures and it was formed by 90 % glycolic acid and 10 % lactic acid. Hydrolysis of PLGA took placed *via* non-enzymatic hydrolysis, and produced lactic and glycolic acids as the final products; these were eliminated from the living organism following a Krebs cycle [261]. Studies have been carried out to determine the degradation time of PLGA by using copolymers with different lactic/glycolic molar ratios [262]. As mentioned above, it was observed that on decreasing the PGA content from 50:50 to 85:15 degradation was delayed from 1-2 to 5-6 months, i. e., a fine tuning of the degradation could be achieved by selecting a given molar ratio.

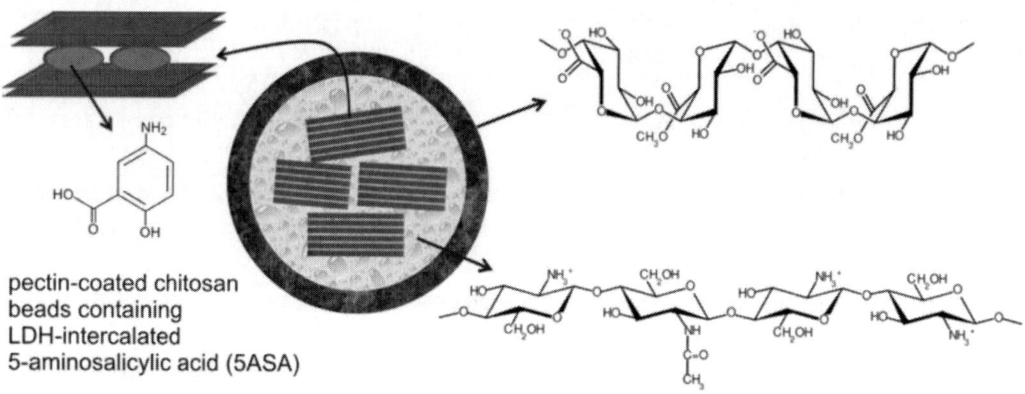

Figure 13. Pectin-coated chitosan bead incorporating 5-aminosalicylic acid (5ASA) intercalated in Mg_2Al layered double hydroxide (LDH) as a new drug delivery system. Reprinted from International Journal of Pharmaceutics, Vol. 463, Ligia N. M. Ribeiro, Ana C. S. Alcântara, Margarita Darder, Pilar Aranda, Fernando M. Araújo-Moreira and Eduardo Ruiz-Hitzky, Pectin-coated chitosan-LDH bionanocomposite beads as potential systems for colon-targeted drug delivery, pp. 1-9. © 2014, with permission from Elsevier.

The studies on systems consisting of drug-intercalated LDHs dispersed in PLGA are very scarce. Chakraborti et al. [263] developed a system to improve the treatment of advanced periodontitis, able to control release of alandronate and tetracycline (TET) to periodontal pocket fluids *via* guided tissue regeneration (GTR) membranes. They first intercalated alandronate in a MgAl-CO3 LDH; the amount of non-intercalated alandronate in the solution was monitored by HPLC with fluorescence detection, and the difference between the total amount of alandronate and that remaining in solution was assumed to be bound to the LDH, well intercalated or merely adsorbed on the external surface of the particles. Two diffraction maxima were recorded at low angle in the PXRD diagram, one at 7.7 Å corresponding to the carbonate-containing LDH, and another at 12.8 Å, suggesting that a partial intercalation of alandronate has taken place, despite the well known affinity of carbonate for LDHs. Release of alandronate was studied in a phosphate buffered saline (PBS) medium at 37 ºC; a fast release (45 %) was observed in the first 6 h, followed by a slow release, reaching a total release value of 80 % after 6 days. It has been suggested that the first, burst process is due to anion exchange close to the particle borders, while the second, slow, release step is limited by diffusion of the drug from inside the interlayer space [264].

Films with 85:15 PLGA were prepared by mixing different compounds and evaporation of the solvent. Release of alandronate was studied from three different systems: (i) one containing the polymer mixed with alandroate; (ii) free alandroate and LDH-CO3 mixed with the polymer and (iii) LDH-alandronate and TET mixed with the polymer.

The release rate of alandronate was strongly influenced by the actual state of alandronate (free or intercalated in the LDH structure) and on the presence of a LDH phase (whichever the intercalated anion, alandronate or carbonate). The burst first release step observed from the LDH alandronate system was practically absent from the composite with the polymer in the presence of a LDH phase, Figure 14. When the drug was dispersed together with MgAl-CO3 in the polymer the release rate decreased; as water molecules entered in the polymer matrix, released alendronate may become dissolved and be re-intercalated in the LDH dispersed in the PLGA matrix, thus accounting for the decrease in the total amount of alandronate released, compared to that observed from films without added LDH.

Release profiles of TET were studied from systems containing, in addition to the polymer, free TET, free TET and LDH-CO3, and free TET and LDH-alandronate. Despite TET was in the "free" (i.e., not intercalated) state in all three cases, its release rate decreased in the presence of both LDH phases, also probably because, as the polymer film was dissolved, TET interacting with the layered particles were eventually intercalated, but no experimental evidence of intercalation was provided.

In all these studies the authors were able to cancel the burst release of alandronate, cancelling its toxic effect and favouring cell growth, allowing the cells to recover and to proliferate.

These same authors have studied [263] similar systems with antibiotics and bactericides; their aim was to study the release of the drugs from drug+polymer and from LDH-drug+polymer mixtures, using a 85:15 PLGA polymer [265]. Release of the drug was monitored by HPLC for TET, sodium fusidate (SF) and vancomycin (VAN), while release of doxorubicin (DOX) and 5-fluorouracil (5FU) was monitored by UV-Vis spectroscopy (at 482 and 254 nm, respectively); finally, release of Antisense oligonucleotide (ASO) was monitored by HPLC and measuring the UV-Vis absorbance at 260 nm. Increasing amounts of drug were added to 2 mg of LDH-CO3, centrifugated and incubated at 37 ºC; the supernatant liquid was analysed for the unintercalated drug, and assuming, on a mass balance basis, that the remaining drug was bound to the LDH particles.

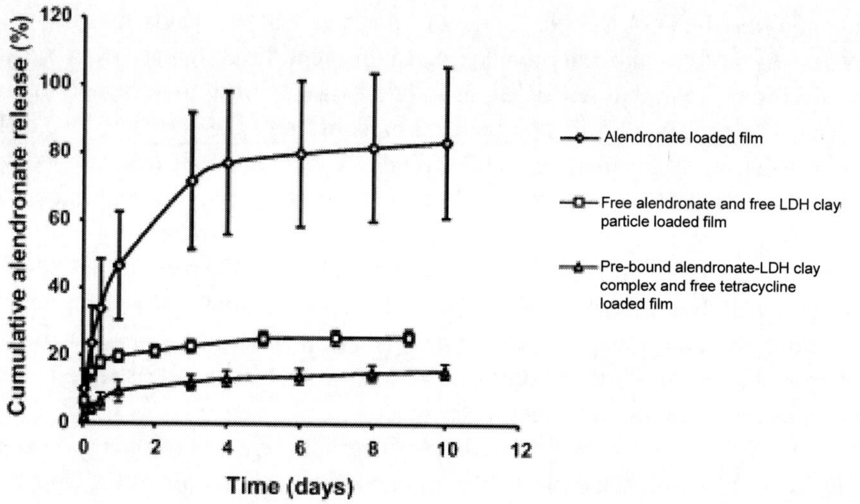

Figure 14. Cumulative release of alendronate from films composed of a polymeric blend of PLGA (85:15) and MePEG (15% w/w) loaded with (%, w/w): (top) 4.2% alendronate, (middle) 5 % tetracycline and 4.2% alendronate-LDH, (bottom) 4.2% alendronate and LDH. All experiments in PBS (0.1 mM), at 37 ºC. Reprinted from International Journal of Pharmaceutics, Vol. 416, Michelle Chakraborti, John K. Jackson, David Plackett, Donald M. Brunette and Helen M. Burt, Drug intercalation in layered double hydroxide clay: Application in the development of nanocomposite film for guided tissue regeneration, pp. 305-313. © 2011, with permission from Elsevier.

As for the former compounds above studied, a first, burst step lasting for ca. 6 h was followed by a slower step for 6 days. However, DOX behaved differently, and only 6 % of the initial amount of drug was released after 6 days. These results can be adapted for some particular clinic applications where a fast elimination of bacteria from infection sensitive sites (e. g., wounds or quirurgical sutures) is required. On the contrary, in other cases a longer exposure to drugs can be desirable for a complete removal of bacteria.

On dispersing LDH-drug (SF, VAN, TET and ASO) systems in the polymer the burst release step was cancelled, obtaining extended release profiles, although only small amounts of drug were released; this might not be a problem, as most of these antibiotic are effective even with low doses. Additionally, degradation of TET (which degradation takes place rather fast in the presence of water when it is bound on the hydrotalcite or on the polymer) is not rapidly degradated even after prolonged incubation in water [266].

Conclusion

We have summarized here the current knowledge on nanocomposites constituted by LDH-hosted drugs dispersed in PLA and other polymers and copolymers. The results confirm the synergic, advantageous role played by this organic-inorganic-organic nanocomposites in drug delivery control. On one side, release of a drug originally dispersed in a polymer can be controlled from the nature of the polymer (chemical composition, particle size and molecular mass, mainly). On the other side, when the drug is hosted in a layered double hydroxide the release can be controlled by the nature of the LDH and its particle size; additionally, the LDH matrix provides a basic medium which can decrease the side-effects of the drug on the living organism. On covering the LDH-drug nanohybrid with a polymer, the former becomes protected from an acidic medium and by selecting the polymer (i. e., its degradation properties), we can even choose the precise place where degradation and drug release will take place. While release from the LDH-drug nanohybrid is usually completed within hours or a few days, the nanocomposites with the polymer permit a long time release, lasting even months, so constituting an alternative procedure for drug delivery.

Further research is however still necessary to optimize the systems, mainly concerning the degradation rate of the polymer and the modification of the LDH to improve the chemical compatibility with the polymer. A combined use of polymers degrading under different experimental conditions (pH medium, mainly) will hopefully permit to direct the nanocomposite to precise targets within the body. In this context, the use of systems with added magnetic components would probably provide an easier method for such a directed delivery, and some few studies on these systems are already available in the literature. Another problem to be addressed concerns the control of the first stages of the drug delivery, which usually lead to a burst liberation of the drug, which is usually undesired.

Acknowledgments

Authors thank finantial support from MICINN (grant MAT2009-08526) and ERDF.

References

[1] Scott, G. Polym. *Degrad. Stab.*, 2000, 68, 1-7.
[2] Mayadunne, R; Adhikari, R. *Biotechnol. Annu. Rev.*, 2006, 12, 301-347.
[3] Pachence, JM; Bohrer, MP; Kohn, J; Robert, L; Joseph, V. In Principles of Tissue Engineering; Ed. Academic: Burlington, 2007, 323-339.
[4] Kiick, KL. *Polymer Therapeutics Sci.*, 2007, 317, 1182-1183.
[5] Tuzlakoglu, K; Reis, RL. *Tissue Eng. Part B Rev.*, 2009, 15, 17-27.
[6] Dash, AK; Cudworth, GC. *J. Pharmacol. Toxicol. Methods*. 1998, 40, 1-12.
[7] Benagiano, G; Gabelnick, H; Farris, M. *Expert Rev. Med. Devices*, 2008, 5, 623-637.
[8] Reinert, KH; Carbone, JP; Sven Erik, J; Brian, F. In Encyclopedia of Ecology; Academic: Oxford, 2008, 3461-3472.
[9] Rives, V; Arco, M del; Martín, C. *J. Control Release*, 2013, 169, 28-39.
[10] Rives, V; Arco, M. del; Martín, C. *Appl. Clay Sci.*, 2014, 88, 239-269.
[11] Ray, SS; Bousmina, M. *Prog. Mater. Sci.*, 2005, 50, 962-1079.
[12] Scheele, CW. *Kgl. Vetenskaps-Academiens nya Handlingar*, 1780, 1, 116-124.
[13] Dobbin, L. The Collected Papers of Carl Wilhelm Scheele, G. Bell & Sons Ltd, London, 1931.
[14] Benninga, H. A History of Lactic Acid Making; Springer: New York, NY, 1990.
[15] Avery, US. Patent 243,827, 1881 (to Avery Lactate Company).
[16] Carothers, WH; Dorough, GL; Van Natta, FJ. *J. Am. Chem. Soc.*, 1932, 54, 761-772.
[17] Gruber, PR; Hall, ES; Kolstad, JJ; Iwen, ML; Benson, RD; Borchardt, RL. U.S. Patent 5,357,035, 1994

[18] Gruber, PR; Hall, ES; Kolstad, JJ; Iwen, ML; Benson, RD; Borchardt, RL. U.S. Patent 6,326,458, 2001.

[19] Bordes, P; Pollet, E; Averous, L. *Prog. Polimers Sci.*, 2009, 34, 125-155.

[20] Vainionpää, S; Rokkoanen, P; Törmälä, P. *Prog. Polym. Sci.*, 1989, 14, 679-716.

[21] www.eis.uva.es/~biopolimeros. *16-04-2009*

[22] Enomoto, K; Ajioka, M; Yamaguchi, A. EP 0572675, 1993 (to Mitsui Chemicals).

[23] Goto, K; Mori, Y; Miyamoto, M. EP 0826711, 1998 (to Mitsui Chemicals).

[24] Ohta, M; Obuchi, S; Yoshida, Y. EP 0603889, 1994 (to Mitsui Chemicals).

[25] Kulkarni, RK; Pani, KC; Neuman, C; Leonard, F. *Arch. Surg.*, 1966, 93, 839-843.

[26] Tsuji, T. Poly(lactic acid) (Japanese),Yoneda Shuppan, Chiba, 2008.

[27] Suzuki, S; Ikada, Y. In Poly(lactic acid): Synthesis, Structures, Properties, Processing, and Applications; Editors Auras, R; Lim, L.-T; Selke, S. E. M; Tsuji, H; Wiley Series on Polymer Engineering and Technology; Wiley: Hoboke, NJ, 2010; chapter 27, 445-456.

[28] Tsuji, H. *Biodegradable Polymers*, Corona Publishing Co. Ltd., Tokyo, 2002.

[29] Tsuji, H; Ikarashi, K. *Biomacromolecules*, 2004, 5, 1021-1028.

[30] Drumright, R; Gruber, P. *Adv. Mater.*, 2000, 23, 1841-1846.

[31] Ray, S; Okamoto, M. *Macromol. Rapid Commun.*, 2003, 24, 815-840.

[32] Wachsen, O; Platkowski, K; Reichert, KH. Polym. Degrad. Stab. 1997, 57, 87-94.

[33] Maeda, H; Miyagawa, H; Yamamoto, K. Jpn. Patent 1999-106554 A, 1999 (to Kobe Steel, Ltd.)

[34] Yu, H; Huang, N; Wang, C; Tang, Z. *J. Appl. Polym. Sci.*, 2003, 88, 2557-2562.

[35] Leiper, HA; McNeill, IC. *Polym. Degrad. Stab.*, 1985, 11, 267-285.

[36] Fan, Y; Nishida, H; Hoshihara, S; Shirai, Y; Tokiwa, Y; Endo, T. *Polym. Degrad. Stab.*, 2003, 79, 547-562.

[37] Kopinke, FD; Mackenzie, K. *J. Anal. Appl. Pyrolysis*, 1997, 40, 43-53.

[38] Cam, D; Marucci, M. *Polymer*, 1997, 38, 1879-1884.

[39] Abe, H; Takahashi, N; Kim, KJ; Mochizuki, M; Doi, Y. *Biomacromolecules*, 2004, 5, 1606-1614.

[40] Quintanar-Guerrero, D; Allemann, E; Fessi, H; Doelker, E. *Drug Dev. Ind. Pharm.*, 1988, 24, 1113-1128.

[41] Sánchez, A; Vila-Jato, JL; Alonso, M. *J. Int. J. Pharm.*, 1993, 99, 263-273.

[42] *Hsu*, Y; Hao, T; Hedley, M. *J. Drug Target,* 1999, 7, 313-323.

[43] Zeng, QH; Wang, DZ; Yu, AB; Lu, GQ. *Nanotechnol.*, 2002, 13, 549-553.

[44] Jaubertie, C; Holgado, MJ; Rives, V; San Román S. *Chem. Mater.*, 2006, 18, 3114-3121.

[45] Losa, C; Marshal-Heussler, L; Orallo, F; Vila Jato, JL; Alonso, M. *J. Pharm. Res.*, 1993, 10, 80-87.

[46] Jong, Y; Jacob, J; Yip, K; Gardner, G; Seitelman, E; Whitney, M; Montgomery, S; *Mathiowitz, E. Nature*, 1997, 386, 410-414.

[47] Gref, R; Minamitake, Y; Peracchia, MT; Trubetskoy, V; Torchilin, V; Langer, R. *Science*, 1994, 263, 1600-1603.

[48] Min, C; Cui, W; Bei, J; Wang, S. *Polym. Adv. Technol.*, 2007, 18, 299-305.

[49] Barakat, I; Dubois, P; Grandfils, C; Jerome, R. *J. Polym. Sci. Part A*, 2001, 39, 294-306.

[50] Koushik, K; Kompella, UB. *Pharm. Res.*, 2004, 21, 524-535.

[51] Malet, P; Odriozola, JA; Labajos, FM; Rives, V; Ulibarri, MA. *Nucl. Instr. Meth. Phys. Res. B*, 1995, 97, 16-19.
[52] Arco, M del; Rives, V; Trujillano, R; Malet, P. *J. Mater. Chem.*, 1996, 6, 1419-1428.
[53] Rives, V; Ulibarri, MA. *Coord. Chem. Rev.*, 1999, 181, 61-120.
[54] Drits, VA; Bookin, AS. In Layered Double Hydroxides: Present and Future; Editor Rives, V; Nova Sci. Pub. Inc.: New York, NY, 2001, Chapter 2, 39-92.
[55] Rives, V., Ed. Layered Double Hydroxides: Present and Future; Nova Sci. Pub. Inc.: New York, NY, 2001, 1-439.
[56] Wypych, F; Satyanarayana, KG. Eds. Clay Surfaces: Fundamentals and Applications, Interface Science and Technology; Elsevier Academic Press: Amsterdam, 2004,Vol. 1, 1-553.
[57] Duan, X; Evans, DG; Eds. *Layered Double Hydroxides; Structure and Bonding*, 119, Verlag: Berlin, 2006, 1-234.
[58] Forano, C; Hibino, T; Leroux, F; Taviot-Guého, C. In Handbook of Clay Science; Eds. Bergaya, F; Theng, BKG; Lagaly, G. Developments in Clay Science; Elsevier: Amsterdam, 2006,Vol. 1, 1021-1095.
[59] Costantino, U; Leroux, F; Nocchetti, M; Mousty, C. In Handbook of Clay Science Part B: Techniques and Applications; Eds. Bergaya, F; Lagaly, G; 2nd Ed; Developments in Clay Science; Elsevier: Amsterdam, 2013, Vol. 5B, 765-791.
[60] Forano, L; Costantino, U; Prévot, V; Taviot-Guého, C. In Handbook of Clay Science Part A: Fundamentals; Eds. Bergaya, F; Lagaly, G; 2nd Ed; Developments in Clay Science; Elsevier: Amsterdam, 2013, Vol. 5A, 745-782.
[61] De la Calle, C; Pons, CH; Roux, J; Rives, *V. Clays Clay Min.*, 2003, 51, 121-132.
[62] Stanimirova, TS; Vergilov, I; Kirov, G; Petrova, N. *J. Mater. Sci.*, 1999, 34, 4153-4161.
[63] Rives, V. In Layered Double Hydroxides: Present and Future; Ed. Rives, V; Nova Sci. Pub. Inc.: New York, NY, 2001, Chapter 4, 115-137.
[64] Hibino, T; Yamashita, Y; Kosuge, K; Tsunashima, A. *Clays Clay Miner.*, 1995, 43, 427-432.
[65] Labajos, FM; Rives, V; Ulibarri, MA. *J. Mater. Sci.*, 1992, 27, 1546-1552.
[66] Rives, V. *Inorg. Chem.*, 1999, 38, 406-407.
[67] Courty, Ph., Marcilly, Ch. In Preparation of Catalysts. III. Scientific Bases for the Preparation of Heterogeneous Catalysts; Eds. Poncelet, G., Grange, P., Jacobs, P. A; Elsevier: Amsterdam, 1983, 485-519.
[68] Serna, CJ; Rendón, JL; Iglesias, JE. *Clays Clay Miner.*, 1982, 10, 180-184.
[69] Hernández-Moreno, MJ; Ulibarri, MA; Rendón, JL; Serna, C. *J. Phys. Chem. Miner.*, 1985, 12, 34-38.
[70] Yamaoka, T; Abe, M; Tsuji, M. *Mater. Res. Bull.*, 1989, 24, 1183-1199.
[71] Reichle, WT. *Solid State Ionics*, 1986, 22, 135-141.
[72] Parise, JB; Hyde, BG. *Acta Cryst. C*, 1986, 42, 1277-1280.
[73] Boehm, HP; Steinle, J; Vieweger, C. Angew. *Chem. Int. Ed. Engl.*, 1977, 16, 265-266.
[74] Clearfield, A. *Chem. Rev.*, 1988, 88, 125-148.
[75] Ma, R; Liu, Z; Takada, K; Iyi, N; Bando, Y; Sasaki, T. *J. Am. Chem. Soc., 2007*, 129, 5257-5263.
[76] Huang, S; Cen, X; Peng, H; Guo, S; Wang, W; Liu, T. *J. Phys. Chem.*, B 2009, 113, 1522-15230.
[77] Miyata, S. *Clays Clay Min.*, 1983, 31, 305-311.

[78] Kooli, F; Jones, W; Rives, V; Ulibarri, MA. *J. Mater. Sci. Lett.*, 1997, 16, 27-29.
[79] Chibwe, K; Jones, W. *J.* Chem. *Soc. Chem. Comm.*, 1989, 926-927.
[80] Kwon, T; Pinnavaia, T. *J. Chem. Mater.*, 1989, 1, 381-383.
[81] Rocha, J; Arco, Mdel; Rives, V; Ulibarri, MA. *J. Mater. Chem.*, 1999, 9, 2499-2503.
[82] Pérez-Ramírez, J; Mul, G; Kapteijn, F; Moulijn, JA. *Mater. Res. Bull.*, 2001, 36, 1767-1775.
[83] Smart, LE; Moore, EA. Solid State Chemistry: An Introduction, 4th Ed., CRC Press: Boca Raton, FL, 2012; 1-465.
[84] Prinetto, F; Ghiotti, G; Graffin, P; Tichit, D. *Microp. Mesop. Mater.*, 2000, 39, 229-247.
[85] Jitianu, M; Zaharescu, M; Balasoiu, M; Jitianu, A. *J. Sol-Gel Sci. Technol.*, 2003, 26, 217-221.
[86] Costantino, U; Marmottini, F; Nocchetti, M; Vivani, R. *Eur. J. Inorg. Chem.*, 1998, 10, 1439-1446.
[87] Adachi-Pagano, M; Forano, C; Bresse, JP. *J. Mater. Chem.*, 2003, 13, 1988-1993.
[88] Rao, MM; Reddy, BR; Jayalakshmi, M; Jaya, VS; Sridhar, B. *Mater. Res. Bull.*, 2005, 40, 347-359.
[89] Hickey, L; Kloprogge, JT; Frost, RL. *J.* Mater. Sci., 2000, 35, 4347-4355.
[90] Benito, P; Labajos, FM; Rocha, J; Rives, V. *Microp. Mesop. Mater.*, 2006, 94, 148-158.
[91] Jacob, J; Chia, LHL; Boey, FYC. *J. Mater. Sci.*, 1995, 6, 5321-5327.
[92] Church, RH. Dielectric properties of low loss minerals. USBOM Report of Investigations 1993, report 9194.
[93] Thostenson, ET; Chou, TW. *Composites A*, 1999, 30, 1055-1071.
[94] Rao, KJ; Vaidhayanathan, B; Ganguli, M; Ramarishanan, PA. *Chem. Mater.*, 1999, 11, 882-895.
[95] Jones, DA; Lelyveld, TP; Mavrofidis, SD; Kingman, SW; Miles, NJ. *Resources Conservation Recycling*, 2002, 34, 75-90.
[96] Chen, CM; Chen, M; Leu, FC; Hsu, SY; Wang, SC; Shi, SC; Chen, CF. *Diamond Related Mater.*, 2004, 13, 1182-1186.
[97] Wathey, B; Tierney, J; Lidströn, P; Westman, *J. Drug Discov. Today*, 2002, 7, 373-380.
[98] Lam, S. S; Russell, A; Lee, CL; Chase, HA. *Fuel*, 2012, 92, 327-339.
[99] D'Arrigo, M.C; Siligardi, C; Leonely, C; So, JY; Kim, HS. *J. Porous Mater.*, 2002, 9, 299-305.
[100] Fetter, G; Hernández, V; Rodríguez, V; Valenzuela, MA; Mara, VH; Bosch, P. *Mater. Lett.*, 2003, 57, 1220-1223.
[101] Kannan, S; Jasra, RV. *J. Mater. Chem.*, 2000, 10, 2311-2314.
[102] Benito, P; Labajos, FM; Rives, V. *Bol. Soc. Esp. Ceram. Vidrio*, 2004, 43, 56-59.
[103] Benito, P; Labajos, FM; Rives, V. In Solid State Chemistry Research Trends, Ed. Bockley, R. W; Nova Sci Pub. Inc.: New York, NY, 2007, 173-225.
[104] Benito, P; Labajos, FM; Rives, V. *Pure Appl. Chem.*, 2009, 81, 1459-1471.
[105] Zhang, X; Hayward, DO; Mingos, MP. *Catal. Lett.*, 2003, 88, 33-38.
[106] Das, NN; Konar, J; Mohanta, MK; Srivastava, SC. *J. Colloid Interface Sci.*, 2004, 270, 1-8.
[107] Hermosín, MC; Pavlovic, I; Ulibarri, MA; Cornejo, J. *J. Environ. Sci. Health A.*, 1993, 28, 1875-1888.

[108] Klumpp, E; Contreras-Ortega, C; Klahre, P; Tino, FJ; Yapar, S; Portillo, C; Stegen, S; Queirolo, F; Schwuger, M. *J.* Coll. *Surf. A: Physicochem. Eng. Aspects*, 2004, 230, 111-116.
[109] Goh, KH; LimTT; Dong, Z. *Water Res.*, 2008, 42, 1343-1368.
[110] Ulibarri, MA; Pavlovic, I; Hermosín, MC; Cornejo, *J. Appl. Clay Sci.*, 2005, 10, 131-145.
[111] Carriazo, D; Arco, M. del, Martín, C; Rives, V. *Appl. Clay Sci.*, 2007, 37, 231-239.
[112] Oespchuck, JM. IEEE Trans. *Microwave Theory Techniques*, 1983, 32, 1200-1224.
[113] Ticot, D; Coq, B. *Cattech*, 2003, 7, 206-217.
[114] Albertazi, S; Bsasile, F; Vaccari, A. In Clay Surfaces: Fundamentals and Applications; Eds. Wypych, F; Satyanarayana, K. G; Elsevier: Amsterdam, 2004; 496-546.
[115] Centi, G; Perathoner, S. *Microp. Mesop. Mater.*, 2008, 107, 3-15.
[116] Rives, V; Carriazo, D; Martín, C. In Pillared Clays and Related Catalysts; Eds. Gil, A; Korili, S.A; Trujillano, R; Vicente, M.A; Springer: New York, N. Y., 2010, 319-397.
[117] Cavani, F; Trifirò, F; Vaccari, A. *Catal. Today*, 1991, 11, 173-301.
[118] Velu, S; Suzuki, K; Okazaki, M; Kapoor, MP; OsakiT; Ohashi, F. *J. Catal.*, 2000, 194, 373-384.
[119] Winter, F; Wolters, M; Van Dillen, AJ; Jong, KP. *Appl. Catal. A: General*, 2006, 307, 231-238.
[120] Castiglioni, GL; Guercio, A; Vaccari, A; Lancia, R. *J. Porous Mater.*, 1995, 2, 79-84.
[121] Medina, F; Dutartre, R; Tichit, D; Coq, B; Dung, NT; Salagre, P; Sueiras, JE. *J. Molec. Catal., A: Chemical*, 1997, 119, 201-207.
[122] Monzón, A; Romeo, E; Royo, C; Trujillano, R; Labajos, FM; Rives, V. *Av. Ing. Quím.*, 1998, 8, 24-27.
[123] Monzón, A; Romeo, E; Royo, C; Trujillano, R; Labajos, FM; Rives, V. *Appl. Catal. A: General*, 1999, 185, 53-63.
[124] Auer, S; Grunwaldt, JD; Köppel, RA; Baiker, A. *J. Molec. Catal. A: Chemical*, 1999, 139, 305-313.
[125] Pinnavaia, TJ; Rameswaran, M; Dimotakis, ED; Giannelis, EP; Righlor, EG. *Faraday Discuss. Chem. Soc.*, 1989, 87, 227-237.
[126] Parashar, P; Sharma, V; Agarwal, DD; Richhariya, N. *Mater. Lett.*, 2012, 74, 93-95.
[127] Xu, ZP; Lu, GQ. *Pure Appl. Chem.*, 2006, 78, 1771-1779.
[128] Ladewig, K; Xu, ZP; Lu, GQ. *Expert Opin. Drug Deliv.*, 2009, 6, 907-922.
[129] Cunha, VRR; Ferreira, AMC; Constantino, RL; Tronto, J; Valim, JB. *Quím. Nova*, 2010, 33, 159-171.
[130] Komarneni, S; Newalkar, BL; Li, D; Tarun, G; Lopano, CL; Heaney; PJ; Post, JE. *J. Porous Mater.*, 2003, 10, 243-248.
[131] Hussein, MZ; Yahaya, A.H; Ping, ML. *Dyes Pigments*, 2004, 63, 135-140.
[132] Pérez Bernal, ME; Ruano Casero, RJ; Rives, V. Ceram. *Silik.*, 2004, 48, 145-154.
[133] Ruano Casero, RJ; Pérez Bernal, ME; Rives, VZ. *Anorg. Allgm. Chem.*, 2005, 631, 2142-2150.
[134] Sánchez, A; Pérez Bernal, ME; Ruano Casero, RJ; Rives, V. In Materiales Arcillosos: de la Geología a las Nuevas Aplicaciones; Eds. Suárez, M; Vicente, M. A; Rives, V; Sánchez, M J; *Varona: Salamanca*, 2006,151-164.
[135] García-García, JM; Pérez-Bernal, ME; Ruano, RJ; Rives, V. *Solid State Sci.*, 2007, 9, 1115-1125.

[136] Pérez-Bernal, ME; Ruano, RJ; Rives, V. *J. Solid State Chem.*, 2009, 182, 2566-2578.
[137] Domínguez, M; Pérez-Bernal, ME; Ruano, RJ; Barriga, C; Rives, V; Ferreira, RS; Carlos, LD; Rocha, *J. Chem. Mater*. 23, 2011, 1993-2004.
[138] Rives, V; Pérez-Bernal, ME; Ruano, RJ; Nebot-Diaz, I. *J. Europ. Ceram. Soc.*, 2012, 32, 975-987.
[139] Rives, V; Pérez-Bernal, ME; Ruano, RJ; Nebot-Diaz, I. In Hydroxides: Synthesis, Types and Applications; Eds. Carillo, A. C; Griego, D. A; Chemical Engineering Methods and Technology; Nova Sci. Pub. Inc.: New York, NY, 2012; 195-220.
[140] Wang, X; Zhang, Q. *Polym. Int.*, 2004, 33, 698-707.
[141] Kalouskova, R; Novota, M; Vymazal, Z. *Polym. Degrad. Stab.*, 2004, 85, 903-909.
[142] Chen, W; Qu, B. *Chem. Mater.*, 2003, 15, 3208-3213.
[143] Pereira, CMC; Herrero, M; Labajos, FM; Marques, AT; Rives, V. *Polym. Degrad. Stab.*, 2009, 94, 939-946.
[144] Camino, G; Maffezzoli, Z; Braglia, M; Lazzaro, M; Zammarano, M. *Polym. Degrad. Stab.*, 2001, 74, 457-464.
[145] Watanabe, T; Aoki, S; Ohta, S; Shirono, K; Tanaka, A. US Patent 5929133, 1997.
[146] Zhao, Y; Li, F; Zhang, R;Evans, DG; Duan, X. *Chem. Mater.*, 2002, 14, 4286-4291.
[147] He, Q; Yin; S; Sato, T. *J. Phys. Chem. Solids*, 2004, 65, 395-402.
[148] Guth, U; Brosda, S; Schomburg, J. *Appl. Clay Sci.,* 1996, 11, 229-324.
[149] Gardner, E; Kristen, MH; Pinnavaia, T. *J. Adv. Mater.*, 2001, 13, 1263-1266.
[150] Fan, Y; Nishida, H; Mori, T; Shirai, Y; Endo, T. *Polymer*, 2004, 45, 1197-1205.
[151] Park, KI; Xanthos, M. *Polym. Degrad. Stab.*, 2009, 94, 834-844
[152] Dagnon, KL; Ambadapadi, S; Shaito, A; Ogbomo, SM; DeLeon, V; Golden, TD; Rahimi, M; NguyeK; Braterman, PS; D'Souza, NA. *J. Appl. Polym. Sci.*, 2009, 113, 1905-1915.
[153] Miao, YE; Zhu, H; Chen, D; Wang, R; Tjiu, WW; Liu, T. *Mater. Chem. Phys.*, 2012, 134, 623-630.
[154] DeLeon, VH; Nguyen, TD; Nar, M; D'Souza, NA; Golden, TA. *Mater. Chem. Phys.*, 2012, 132, 409-415.
[155] Yang, JH; Han, YS; Park, M; Park, T; Hwang, SJ; Choy, JH. *Chem. Mater.*, 2007, 19, 2679-2685.
[156] Gu, Z; Thomas, AC; Xu, ZP; Campbell, JH; Lu, GQ. *Chem. Mater.*, 2008, 20, 3715-3722.
[157] Zhang, X; Xue, L; Wang, J; Liu, Q; Liu, J; Gao, Z; Yang, W. *Colloids Surfaces A,* 2013, 431, 80-86.
[158] Alonso, MJ; Cohen, S; Park, TG; Gupta, RK; Siber, GR; Langer, R. *Pharm. Res.*, 1993, 10, 945-953.
[159] Zolnik, BS; Burgess, DJ. *J. Control. Release*, 2007, 122, 338-344.
[160] Hu, C; Liu, S; Zhang, Y; Li, B; Yang, H; Fan, C; Cui, W. *Acta Biomater.*, 2013, 3, 7381-7388.
[161] Zhou, J; Han, S; Dou, Y; Lu, J; Wang, C; He, H; Li, X; Zhang, J. *Int. J. Pharm.*, 2013, 448, 175-188.
[162] Montes, A; Kin, N; Gordillo, MD; Pereyra, C; Martínez de la Ossa, EJ. *J. Supercrit. Fluids*, 2014, 89, 58-67.
[163] Takahashi M; Onishi H; Machida Y. *J. Control. Release*, 2004, 100, 63-74.
[164] Onishi, H; Takahashi, M; Machida, Y. *Biol. Pharm. Bull.*, 2005, 28, 2011-2015.

[165] Corrigan, OI; Li, X. *Eur. J. Pharm. Sci.*, 2009, 37, 477-485.
[166] Zakeri-Milani, P; Delf Loveymi, B; Jelvehgari, M; Valizadeh, H. *Colloids Surf. B Biointerfaces*, 2013, 103, 174-181.
[167] Hami, Z; Amini, M; Ghazi-Khansari, M; Rezayat, SM; Gilani, K. *Colloids Surf. B Biointerfaces*, 2014, 116, 309-317.
[168] Bédouet, L; Moine, L; Pascale, F; Nguyen, VN; Labarre, D; Laurent, A. *Int. J. Pharm.*, 2014, 459, 51-61.
[169] Zhang, K; Tang, X; Zhang, J; Lu, W; Lin, X; Zhang, Y; Tian, B; Yang, H; He, H. *J. Control. Release*, 2014, 183, 77-86.
[170] Yoo, HS; Park, TG. *J. Control. Release*, 2001, 70, 63-70.
[171] Kim, YJ; Choi, S; Koh, JJ; Lee, M; Ko, KS; Kim, SW. *Pharm. Res.*, 2001, 18, 548-550.
[172] Duan, Y; Zhang, Y; Gong, T; Zhang, Z. *J. Mater. Sci. Mater. Med.*, 2007, 18, 2067-2073.
[173] Li, S; McCarthy, S. *Biomaterials*, 1999, 20, 35-44.
[174] Zabihi, F; Xin, N; Li, S; Jia, J; Cheng, T; Zhao, Y. *J. Supercrit. Fluids*, 2014, 89, 99-105.
[175] Thote, AJ; Gupta, RB. *Nanomed. Nanotechnol.*, 2005, 1, 85-90.
[176] Costantino, U; Nocchetti, M. In Layered Double Hydroxides: Present and Future; Ed. Rives, V; Nova Sci. Pub. Inc.: New York, NY, 2001, 383-411.
[177] Khan, AI; Lei, L; Norquist, AJ; O´Hare, D. *Chem. Comm.*, 2001, 2342-2343.
[178] Choy, JH; Choi, SJ; Oh, JM; Park, T. *Appl. Clay Sci.,* 2007, 36, 122-132.
[179] Costantino, U; Nocchetti, M; Sisani, M; Vivani, RZ. *Kristallogr.*, 2009, 224, 273-281.
[180] Jakubiková, B; Kovanda, F. *Chemicke Listy*, 2010, 104, 906-912.
[181] Chakraborty, M; Bose, P; Mandal, T; Data, B; Tarak, D; Shilpa, P; Chakaborty, J; Manoj, M; Basu, D. *Trans. Ind. Ceram. Soc.*, 2010, 69, 229-234.
[182] Chakraborty, M; Dasgupta, S; Sengupta, S; Chakraborty, J; Basu, D. *Trans. Ind. Ceram. Soc.*, 2010, 69, 153-163.
[183] Wang, Q; O'Hare, D. *Chem. Rev.*, 2012, 112, 4124-4155.
[184] Riaz, U; Ashaf, SM. *Minirev. Med. Chem.*, 2013, 13, 522-529.
[185] Schmassmann, A; Tarnawski, A; Gerber, HA; Flogerzi, F; Sanner, M; Varga, L; Halter, F. *Gut.*, 1994, 35, 896-904.
[186] Van der Voet, GB; Wolf, FA. *Clin. Toxicol.*, 1896, 24, 545-553.
[187] Ookubo, A; Ooi, K; Hayashi, H. *J. Pharm. Sci.*, 1992, 81, 1139-1140.
[188] O'Hare, DM. European Patent 1341556 (2002).
[189] Ambrogi, V; Fardella, G; Grandolini, G; Nocchetti, M; Perioli, L. *J. Pharm. Sci.*, 2003, 92, 1407-1418.
[190] Arco, Mdel; Cebadera, E; Gutiérrez, S; Martín, C; Montero, MJ; Rives, V; Rocha, J; Sevilla, MA. *J. Pharm. Sci.*, 2004, 93 1649-1658.
[191] Perioli, L; Pagano, C. *Drug. Deliv.*, 2012, 9, 1559-1572.
[192] Perioli, L; Mutascio, P; Pagano, C. *Pharm. Res.*, 2013, 30, 156-166.
[193] Ay, AN; Zümreoglu-Karan, B; Temel, A; Rives, V. *Inorg. Chem.*, 2009, 48, 8871-8877.
[194] Nguyen, QT., Baird, DG. *Adv. Polym. Technol.*, 2006, 25, 270-285.
[195] Novak, BM. *Adv. Mater.*, 2003, 422-433.
[196] Utracki, LA. Clay-Containing Polymeric Nanocomposites, Rapra Technology Ltd.: Shawbury, 2001, Vol 1-2, 1-755.
[197] Utracki, LA; Sepehr, M; Boccaleri, E. *Polym. Adv. Technol.*, 2007, 18, 1-37.

[198] Swanson, CH; Stimpfling, T; Troutier-Thulliez, AL; Hintze-Bruening, H; Leroux, F. *J. Appl. Polym. Sci.*, 2013, 128, 2954-2960.
[199] Alexandre, M; Dubois, P. *Mater. Sci. Eng.*, 2000, 28, 1-63.
[200] Kornmann, X. Synthesis and Characterization of Thermoset-Clay Nanocomposites, Ph. D. Thesis, Division of Polymer Engineering, Lulea University of Technology, Lulea, Sweden, 2001.
[201] Gu, SY; Ren, J; Dong, B. *J. Polym. Sci. B*, 2007, 45, 3189-3196.
[202] Vidotti, SE; Chinellato, AC; Hu, GH; Pessan, LA. *J. Polym. Sci.*, B 2007, 45, 3084-3091.
[203] Wang, GA; Wang, CC; Chen, CY. *Polymer*, 2005, 46, 5065-5074.
[204] Costa, FR; Leuteritz, A; Wagenknecht, U; Jehnichen, D; Häussler, L; Heinrich, G. *Appl. Clay Sci.,* 2008, 38, 153-164.
[205] Qiu, L; Chen, W; Qu, B. *Polymer*, 2006, 47, 922-930.
[206] Gao, F. *Mater. Today*, 2004, 7, 50-55.
[207] Usuaki, A; Kojima, Y; Kawasumi, M; Okada, A; Fukushima, A; Kurauchi, T; Kamigaito, O. *J. Mater. Res.*, 1993, 8, 1179-1184.
[208] Chen, W; Qu, BJ. *J. Mater. Chem.*, 2004, 14, 1705-1710.
[209] Haggenmueller, R; Du, FM; Fishcer, JE; Winey, KI. *Polymer*, 2006, 47, 2381-2388.
[210] Vaia, RA; Ishii, H; Giannelis, EP. *Chem. Mater.*, 1993, 5, 1694-1696.
[211] Zhao, N; Shi, S; Lu, G; Wei, M. *J. Phys. Chem. Solids*, 2008, 69, 1564-1568.
[212] Pan, P; Zhu, B; Dong, T; Inoue, Y. *J. Polym. Sci. B: Polym. Phys.*, 2008, 46, 2222-2233.
[213] Xu, ZP; Braterman, PS. *J. Mater. Chem.*, 2003, 13, 268-273.
[214] Ha, JU; Xanthos, M. *Appl. Clay Sci.,* 2010, 47, 303-310.
[215] Chiang, MF; Wu, TM. *Composites Sci. Technol.*, 2010, 70, 110-115.
[216] Chiang, MF; Chu, MZ; Wu, TM. *Polymer Degr. Stability*, 2011, 96, 60-66.
[217] Katiyar, V; Gerds, N; Koch, CB; Risbo, J; Hansen, HCB; Plackett, D. *Polymer Degr. Stability*, 2010, 95, 2563-2573.
[218] Wang, DY; Leuteritz, A; Wang, YZ; Wagenknecht, U; Heinrich, G. *Polymer Degr. Stability*, 2010, 95, 2474-2480.
[219] Gerds, N; Katiyar, V; Koch, CB; Hansen, HCB; Plackett, D; Larsen, EH; Risbo, J. *Polymer Degr. Stability*, 2012, 97, 2002-2009.
[220] Schmidt, B; Katiyar, V; Plackett, D; Larsen, EH; Gerds, N; Koch, CB; Petersen, JH. *Food Addit. Contam. Part A, Anal. Control Expo. Risk Assess.*, 2011, 28, 956-966.
[221] Eili, M; Shameli, K; Ibrahim, NA; Yunus, ZW. *Int. J. Mol. Sci.*, 2012, 13, 7938-7951.
[222] Bouissou, C; Rouse, J; Price, R; Walle, C. *Pharm. Res.*, 2006, 23 1295-1305.
[223] Wu, M; Kleiner, L; Tang, FW; Hossainy, S; Davies, MC; Roberts, CJ. *Eur. J. Pharm. Sci.*, 2009, 36, 493-501.
[224] Pan, D; Zhang, H; Zhang, T; Duan, X. *Chem. Eng. Sci.*, 2010, 65, 3762-3771.
[225] San Román, MS; Holgado, MJ; Salinas, B; Rives, V. *Appl. Clay Sci.,* 2013, 71, 1-7.
[226] San Román, MS; Holgado, MJ; Salinas, B; Rives, V. *Appl. Clay Sci.,* 2012, 55, 158-163.
[227] Ha, JU; Xanthos, M. *Int. J. Pharm.*, 2011, 414, 321-331.
[228] Bugatti, V; Costantino, U; Gorrasi, G; Nocchetti, M; Tammaro, L; Vittoria, V. *Eur. Polym. J.*, 2010, 46, 418-427.

[229] Costantino, U; Ambrogi, V; Nocchetti, M; Periolo, L. *Micropor. Mesopor. Mat.*, 2007, 107, 149-160.

[230] Tammaro, L; Tortora, M; Vittoria, V; Costantino, U; Marmottini, F. *J. Polym. Sci. A: Polym. Chem.*, 2005, 43, 2281-2290.

[231] Okada, A; Usuki, A. *Mater. Sci.* Eng. C 1995, 3, 109-115.

[232] Tammaro, L; Costantino, U; Bolognese, A; Sammartino, G; Marenzi, G; Calignanoe, A; Tetè, S; Mastrangelo, F; Califano, L; Vittoria, V. *Int. J. Antimicrob. Agents*, 2007, 29, 417-423.

[233] Sammartino, G; Marenzi, G; Tammaro, L. *Int. J. Immunopathol. Pharmacol.*, 2006, 19, 53-58.

[234] Tammaro, L; Costantino, U; Nocchetti, M; Vittoria, V. *Appl. Clay Sci.*, 2009, 43, 350-356.

[235] Costantino, U; Montanari, F; Nocchetti, M; Canepa, F; Frache, A. *J. Mater. Chem.*, 2007, 17, 1079-1086.

[236] Costantino, U; Bugatti, V; Gorrasi, G; Montanari, F; Nocchetti, M; Tammaro, L; Vittoria, V. *Appl. Mater. Interfaces*, 2009, 1, 668-677.

[237] Triantafillidis, CS; LeBaron, PC; Pinnavaia, TJ. *J. Solid State Chem.*, 2002, 167, 354-362.

[238] Bugatti, V; Gorrasi, G; Montanari, F; Nocchetti, M; Tammaro, L; Vittoria, V. *Appl. Clay Sci.*, 2011, 52, 34-40.

[239] Kalinichev, AG; Kumar, PP; Kirkpatrick, R. *J. Philos. Mag.*, 2010, 90, 2475-2488.

[240] Romeo, V; Gorrasi, G; Vittoria, V; Chronakis, IS. *Biomacromolecules*, 2007, 8, 3147-3152.

[241] Valarezo, E; Tammaro, L; González, S; Malagón, O; Vittoria, V. *Appl. Clay Sci.*, 2013, 72, 104-109.

[242] Ambrogi, V; Perioli, L; Ricci, M; Pulcini, L; Nocchetti, M; Giovagnoli, S; Rossi, C. *Micropor. Mesopor. Mat.*, 2008, 115, 405-415.

[243] Lorenzo-Lamosa, ML; Remunan-Lopez, C; Vita-Joto, YL; Alonso, MJ. *J. Control. Release*, 1998, 52, 109-118.

[244] Choi, G; Lee, JH; Oh, YJ; Choy, YB; Park, MC; Chang, HC; Choy, JH. *Int. J. Pharm.*, 2010, 402, 117-122.

[245] Jung, H; Kim, HM; Choy, YB; Hwang, SJ; Choy, JH. *Int. J. Pharm.*, 2008, 349, 283-290.

[246] Park, JK; Choy, YB; Oh, JM; Kim, JY; Hwang, SJ; Choy, JH. *Int. J. Pharm.*, 2008, 359, 198-204.

[247] Jung, H., Kim, HM., Choy, YB., Hwang, SJ., Choy, JH. *Appl. Clay Sci.*, 2008, 40, 99–107.

[248] Ha, JU; Xanthos, M. *Polym. Compos.*, 2009, 30, 534-542.

[249] Arco, Mdel; Fernández, A; Martín, C; Rives, V. *J. Solid State Chem.*, 2010, 183, 3002-3009.

[250] Li, B; He, J; Evans, DG; Duan, X. *Appl. Clay Sci.*, 2004, 27, 199-207.

[251] Li, B; He, J; Evans, DG; Duan, X. *Int. J. Pharm.*, 2004, 287, 89-95.

[252] Wang, L; Li, B; Hu, Z; Cao, J. *Appl. Clay Sci.*, 2013, 72, 138-146.

[253] Xu, K; Chen, G; Shen, J. *Appl. Clay Sci.*, 2013, 75, 114-119.

[254] Liu, Y; Song, J; Jiao, F; Huang, J. *J. Mol. Struct.*, 2014, 1064, 100-106.

[255] Decher, G. *J. Sci.*, 1997, 277, 1232-1237.

[256] Drezdzon, MA. *Inorg. Chem.*, 1988, 27, 4628-4632.
[257] Ribeiro, LNM; Alcântara, ACS; Darder, M; Aranda, P; Araújo-Moreira, FM; Ruiz-Hitzky, E. *Int. J. Pharm.*, 2014, 463, 1-9.
[258] Zhang, H; Zou, K; Sun, H; Duan, X. *J. Solid State Chem.*, 2005, 178, 3485-3493.
[259] Sharma, R; Ahuja, M. *Carbohyd. Polym.*, 2011, 85, 658-663.
[260] Shen WX. In Encyclopedic Handbook of Biomaterials and Bioengineering; Eds. Wise, L. M; Trantolo, D. J; Altobelli, D. E; Yaszemski, M. J; Gresser, J. D; Schwartz, E. R; Marcel Dekker : New York, NY, 1995, 1151-2000.
[261] Lewis, DH. In Biodegradable Polymers as Drug Delivery Systems; Eds. Chasin, M;. Langer, R; Drugs and the Pharmaceutical Sciences; Marcel Dekker: New York, NY, 1990; 1-41.
[262] Thomasin, C; Corradin, G; Men, Y; Merkle, HP; Gander, B. *J. Control Release*, 1996, 41, 131-145.
[263] Chakraborti, M; Jackson, JK; Plackett, D; Brunette, DM; Burt, HM. *Int. J. Pharm.*, 2011, 416, 305-313.
[264] Ambrogi, V; Fardella, G; Grandolini, G; Perioli, L. *Int. J. Pharm.*, 2001, 220, 23-32.
[265] Chakraborti, M; Jackson, JK; Plackett, D; Gilchrist, SE; Burt, HM. *J. Mater. Sci: Mater. Med.*, 2012, 23, 1705-1713.
[266] Ryu, SJ; Jung, H; Oh, JM; Lee, JK; Choy, JH. *J. Phys. Chem. Solids*, 2010, 71, 685-688.

In: New Developments in Polylactic Acid Research
Editor: Courtney Winthrop
ISBN: 978-1-63463-054-2

Chapter 3

PLA Enhanced via Plasma Technology: A Review

Pieter Cools, Nathalie De Geyter and Rino Morent
Research Unit Plasma Technology, Department of Applied Physics,
Ghent University, Ghent, Belgium

Abstract

In the last 50 years, it has numerously been evidenced that plasma treatments can effectively change the surface properties of conventional polymers (PP, PET, PE, PA…). Only recently the influence of plasmas on biodegradable polymers, such as polylactic acid (PLA) has been reported. PLA can have important biomedical applications in tissue engineering as three-dimensional porous structures (scaffolds). But their low surface energy leads to poor cell attachment and proliferation and this limits the success of these biodegradable scaffolds. The response it elicits from the surrounding biological environment is crucial and this response is mainly governed by the scaffold surface properties. Therefore, PLA surface properties need to be modified to introduce additional functional groups, which can be recognized as adhesion sites for surrounding cells. Different methods have been developed to obtain the wanted surface properties, however, in the past decade, the use of non-equilibrium plasmas for selective surface modification has been a rapidly growing research field. This chapter therefore presents a critical overview on recent advances in plasma-assisted surface modification of PLA.

1. Introduction

Amongst other biodegradable materials, PLA can have important biomedical applications in tissue engineering as three-dimensional porous structures (scaffolds). On the one hand it exhibits excellent mechanical properties and a relatively fast degradation rate, making it an excellent choice of material. On the other hand it suffers from a low surface free energy, leading to a reduced cell adhesion and proliferation, drastically limiting its use as a scaffold material. The response it elicits from the surrounding biological environment is crucial and this response is mainly governed by the scaffold surface properties.

In the last few decades, numerous surface modification techniques have been developed that introduce extra functionalities on the substrate surface, leading to an increase in surface free energy. When applying a surface modification in general, it is critical that bulk properties, such as mechanical strength and structural integrity, remain unaffected. When it comes to biomedical applications specifically, the use of solvents and chemicals based surface treatment techniques are reduced to a limited set approved by EU and US legislation. Therefore solvent free techniques such as γ-radiation, UV treatments, corona discharges etc. have gained a lot of popularity in the field of tissue engineering. One of those techniques that has been around for a long time has recently found its way into the biomedical field: non-thermal plasma technology.

In the last 50 years, it has numerously been evidenced that non-thermal plasma technology can effectively change the surface properties of conventional polymers (PP, PET, PE, etc.) [1-5]. Although biodegradable polymers are already commercially available for more than 35 years, the studies on plasma modification of these materials situate themselves more recently. The first publications started around the period Langer and Vacanti introduced the concept of tissue engineering and their number has kept steadily growing ever since [6].

At the end of last century, literature was dominated by oxygen plasma treatments in low pressure RF discharges and the corresponding response on cell adhesion, growth and proliferation. Although today the majority of publications still focuses on these low pressure oxygen plasma, there is a growing field around atmospheric pressure plasma treatments. Both topics, as well as other discharge gasses, plasma polymerization, grafting, biomedical applications and applications outside the biomedical field will be discussed in the following paragraphs.

2. Non-Thermal Plasma Technology

2.1. History and Definitions

Plasma is a gaseous mixture of radicals, ions, electrons and neutrals that is also known as the fourth state of matter and was first defined by Langmuir in 1929. Over 99% of the known universe consists of plasma, earth being one of the few exceptions. In nature, plasma occurs under the form of lighting or aura borealis (northern lights), but the majority of plasma on earth are man-made. In the lab plasmas are normally generated by supplying energy to a neutral gas, via thermal energy (flames) or by applying an electrical field causing the formation of charge carriers [7].

An important distinction has to be made between thermal and non-thermal plasma. In thermal plasma, both electrons and heavy particles have the same temperature situated around 4000-5000 K or higher and are considered to be in a thermal equilibrium. For non-thermal plasma, only the electrons are accelerated by the electrical field, causing a thermal difference between the light and heavy particles, resulting in plasma that is operating at lower temperatures. Due to this difference in operating temperature, they are often referred to as 'hot' and 'cold' plasma respectively. For the cold plasma this still can result in a temperature of a few 1000 K. For biomedical applications, only non-thermal plasma treatments are applied where the degree of ionization is 1% or lower, resulting in plasma operating at room temperature range (290-330 K) thus avoiding thermal degradation of the (biodegradable) polymers. In the field of non-thermal plasma there is still a wide variety in the way plasma can be generated.

2.2. Plasma Sources

Each of the sources discussed in the following section has its advantages and disadvantages. Over time they have found applications in all branches of the industry: automotive, packaging, textiles, aerospace, (bio)medical etc. The plasma reactor designs that have been developed are numerous and a whole book could be written on that topic alone, as for each new applications design changes are made to optimize the plasma treatment for that specific process. But almost all of these designs can be linked to one of the plasma sources discussed here.

DC Discharge

Non-thermal plasma generated via a DC discharge are normally formed in a closed reactor between electrodes at low pressures (10^{-1} – 10 pa) [7, 8]. Depending on the used voltage and current, different types of discharges can be obtained (Figure 1). The Townsend discharge is a self-sustaining discharge, typically characterized by a low current. An increase in current results in a drop of voltage and forms a glow discharge. It is this current-voltage region that is normally applied for the modification of polymers, as it guarantees a homogeneous treatment al throughout the reactor.

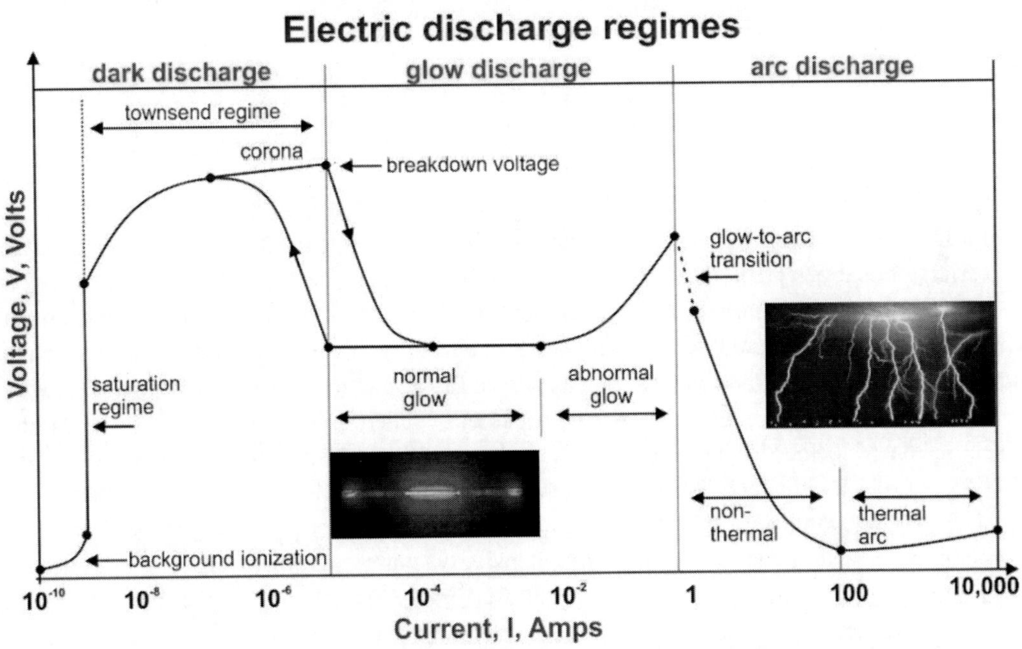

Figure 1. An overview of the different electric discharge regimes possible [86].

A further increase of the discharge current results in the build-up of the corresponding voltage until an arc is formed and the voltage drops almost completely. In general it can be concluded that all processes going on in a DC discharge are well-known, giving a high control over the process.

Next to the continuous DC discharge, also pulsed systems are available. Two of the main advantages of pulsed systems, concerning biomedical applications, are the possibility of using higher discharge powers without thermally damaging the substrates and if used for the deposition of thin films, it applies a more homogeneous coating.

Alongside the limited pressure range, a major disadvantage of the DC discharge technology is the direct exposure of the metal electrodes to the plasma medium, making them vulnerable to corrosion when exposed to certain compounds formed in the plasma.

RF and Microwave Discharges

Compared to DC discharges, radiofrequency and microwave discharges are formed and sustained using high frequency electromagnetic fields [7-9]. RF discharges can operate at a range between 1 – 100 MHz, but in most cases a fixed frequency of 13.56 MHz is applied. Concerning the operating pressure, a wider range (1-10^3 Pa) is possible, but with the exception of a few, high-vacuum equipment is needed, which is expensive and drastically increases treatment times. For biomedical purposes, this is probably the most applied discharge, as it is the plasma technique of choice for oxygen plasma treatments and several reactors are commercially available on the market.

For microwave discharges, the frequency range is an order of magnitude higher and is normally fixed at 2.45 GHz. In contrast to RF and DC discharges, microwave reactors can be operated in a very broad pressure range starting from 1 Pa up to atmospheric pressure. Operating at higher pressures does result in an increase of heat transfer to the substrate, limiting the practical pressure range for biomedical applications, giving cause to the same limitations for the vacuum equipment as for the RF discharge.

Dielectric Barrier Discharge

A dielectric barrier discharge (DBD) or silent discharge is using high frequency AC or RF as a discharge source and together with its high pressure operating range (5-100 kPa), it differs somewhat from the previous described techniques [7, 8]. Already in 1857, Siemens used a DBD for the generation of ozone and to this day it is still one of its most important industrial applications [10].

A DBD reactor typically consists of 2 electrodes, which are usually a few mm apart, of which at least 1 is covered with a dielectric material such as glass, quartz or a ceramic material. The voltage applied ranges typically from 0.5 kV up to a few 100 kV. The discharge formed consists out of numerous small streamers, called micro-discharges. Due to the dielectric material, the discharge current is limited, giving cause to very short-lived micro-discharges (1-10 ns) that are distributing themselves homogeneously across the electrode.

The major advantage of the DBD compared to other plasma techniques, is the possibility to operate at pressures that do not require extensive vacuum equipment, making it a cheaper and more time-effective. Furthermore they allow for a wider range of applications such as plasma chemistry, polymerization, etching, cleaning etc. due to their very low heat generation. Many of these applications are not always possible using other techniques as will be discussed in the next paragraph.

Atmospheric Pressure Plasma Jets

In the previous sections, the most important power supplies have been discussed with their pros and cons and as has been mentioned, they can be used for any kind of reactor design. In this paragraph some special attention will be given to a kind of set-up that has hugely gained in popularity for the last decade: the atmospheric pressure plasma jet (APPJ).

The atmospheric pressure plasma jet (APPJ) is typically a variation of the parallel plate set-up. Through 2 concentric electrodes, whether or not covered with a dielectric material, a mixture of different gasses is send. The high voltage is applied and the ionized gas is send through a nozzle, forming a plasma plume. The substrate is typically placed a few mm underneath the nozzle, in the so-called afterglow, where it is exposed to the reactive species of the plasma [8].

In 2007 Laroussi et al., wrote an excellent review on the topic of arc-free APPJ's that covers all the essentials [11]. Discussing each of them here would lead this chapter too far out of its scope and would not add any crucial information to understand its applicability for biomedical research, which will be discussed in chapter part 4.2.

2.3. Plasma Material Interactions

Before discussing the applications of plasma technology in the biomedical field, it is of the essence to understand the possible interactions of the ions, radicals and electrons present in the plasma and the substrate exposed to them.

Plasma Cleaning and Etching

Contamination is everywhere. During the production process and storage of (bio)materials, they are exposed to a number of solvents, greases, volatiles etc. These products often adsorb on the material surface, resulting in a reduced product performance. A known example in the biomedical field is the adsorption of low molecular weight carbon species onto a pristine titanium sample, when exposed to ambient air. When used as an implant material, this surface pollution results in a reduced cell adhesion, proliferation and growth and in some cases even results in cell death [12, 13].

When exposing these contaminated surfaces to the mixture of highly reactive species generated by the plasma, they remove this contamination in a matter of seconds, without having any major influence on the underlying surface [14]. When further increasing the treatment time and/or discharge power, the plasma is capable of not only cleaning the surface, but also of etching away the top layers of the sample [15-18]. For hard materials, higher discharge powers and/or a prolonged exposure are essential to obtain a notable etching effect. Polymer surfaces on the other hand, are inevitably etched when exposed to non-thermal plasma. For biomedical applications this is considered a benign effect, as possible changes in nano-roughness can amplify the positive effects of plasma treatments concerning cell growth and adhesion [19, 20].

Plasma Activation

The main reason for exposing (biodegradable) polymers, such as PLA, to plasma, is to enhance its surface properties. The charged particles in the plasma interact with the substrate in such a way that reactive radical sites are generated. Depending on the gas feed, these sites will react (in)directly with O/N di-radicals, forming a broad variety of functional groups on the surface. This incorporation of polar groups, as evidenced via x-ray photoelectron spectroscopy (XPS), result in a drastic change in wettability, which in turn has a positive effect on material –cell interactions.

The traditional use of plasma treatments on biodegradable materials is not always considered the right course of action. For some applications such as the inside of artificial stents or heart valves an increase in cell adhesion has to be avoided at all cost, as it will lead to a premature failure of the implant. Instead of using the traditional gas feeds for plasma treatment (noble gasses, oxygen, dry air, nitrogen…) different research groups have used fluorinated gasses such as CF_4 that result in the formation of super hydrophobic surfaces with water contact angles of 150° and higher. These fluorinated surfaces prevent cells from effectively adhering on the surface and thus guaranteeing an optimal performance of the implant material [4, 21, 22].

When plotting the static water contact angle (WCA) as a function of plasma treatment time, while fixating all other parameters, a graph is generated that depicts the treatment effectiveness (see figure 2). In most cases a plasma treatment results in a progressive decrease/increase of the WCA followed by a saturation plateau. Using these plots allows a fast and easy fine-tuning of the surface free energy needed for a specific application.

One of the advantages of plasma activation compared to other techniques introducing functional groups on the surface is that it falls under the category of non-invasive techniques, only modifying the outermost surface layers. This guarantees the preservation of the structural integrity and chemical and mechanical properties of the bulk.

Plasma Grafting and Polymerization

An alternative way of using non-thermal plasma is as an initiation medium for the polymerization of thin films. Almost always preceded by plasma activation, the polymerization process can follow two pathways: plasma grafting is very much the same as radical polymer grafting in traditional chemistry. The radicals introduced on the surface by plasma activation are used as initiation sites to start the chain reaction. Before the monomer is introduced into the plasma chamber, the plasma is deactivated. Therefore the functionalities present in the monomer are preserved and a "traditional" polymer chain is grown.

For plasma polymerization this is not the same case. This time the plasma is used as an initiation medium and remains active during the polymerization reaction. The active radicals present in the plasma form initiation sites on both the substrate surface and the monomer molecules that are mixed in the feed. In contrast to chemical initiation, plasmas are not as specific as to where the radicals are introduced, using the functional groups of the polymer precursor as well to initiate the chain reaction, resulting in a highly cross-linked, pinhole free and completely amorphous thin film that significantly differs from its traditional counterpart and adheres to almost any surface. Varying the discharge power gives a high control over the amount of functionalities preserved in the film. From a biomedical viewpoint this is an interesting application, as the functional group density plays a critical role in the growth and proliferation of cells and differs for the type of cells used.

2.4. Surface Characterization Techniques

As mentioned in section 2.3, when applying non-thermal plasma to a biodegradable polymer such as PLA, a number of changes are induced onto the material surface. In order to understand material-cell interactions, it is critical to thoroughly characterise the induced surface changes. To do so, a wide variety of analysing techniques are available.

Contact Angle Goniometry

The usual strategy is to start with a contact angle goniometry study to map the changes in wettability and surface energy. This is the logical choice to commence with, as it is both an extremely fast technique and gives very sensitive results. For the changes in wettability a droplet of water is deposited onto the surface and the angle between the droplet and the surface is measured (see figure 2). To determine the changes in surface free energy, multiple liquids (usually water and/or diodomethane) are deposited onto the surface and the contact angles are determined. These results are then combined in the harmonic mean equations and the surface free energy is calculated [23]. These results often give a first indication what kind of changes the plasma treatment induces and to what extent, but it gives no chemical information.

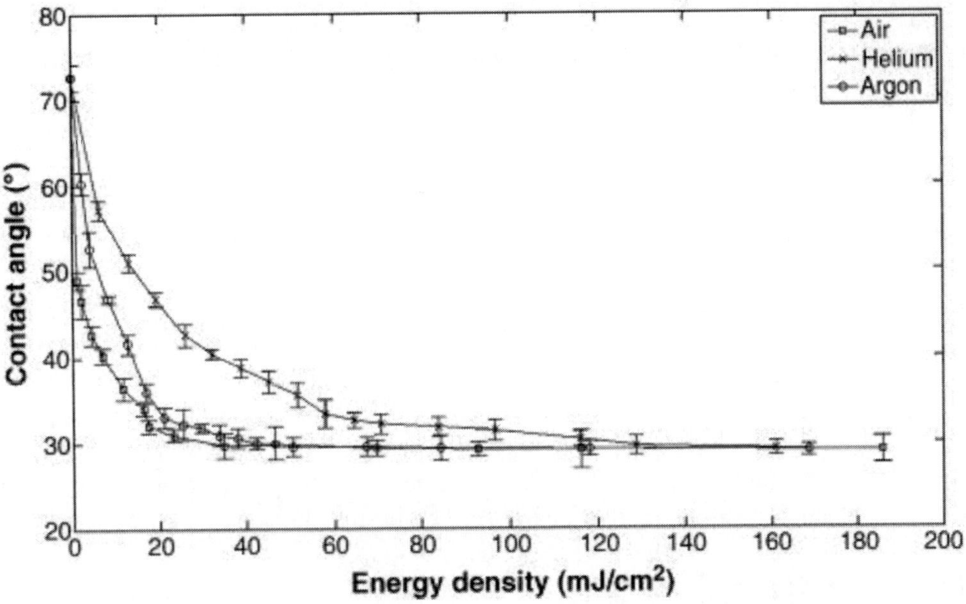

Figure 2. Example of changes in water contact angle as a function of energy density for different discharge gasses (measurements performed on PET, using a DBD discharge at medium pressure) [87].

X-Ray Photoelectron Spectroscopy

The method of choice for a surface chemical analysis is XPS. An X-ray beam with a fixed energy radiates the surface. This excites the inner shells of the atoms and photons with an energy characteristic for those elements are emitted. Via a hemispherical analyser the photons of a certain energy can be focused and collected. Scanning all energies in one spectrum gives you information on the elemental composition of the substrate. The real power of the technique lies in the fact that chemical bonds cause a shift in the energy at which the photon is emitted. This allows obtaining information about the chemical bonds present in the material analyzed via the deconvolution of the measured peaks. One of the major reasons XPS is preferred over other similar techniques is because of the analysis depth (+/-10 nm) lying in the same range as the plasma penetration depth [24]. The downside would be that it only gives information about the chemical bonds present on the surface, making it difficult to distinct certain sets of functional groups. In some cases derivatisation techniques can solve this problem, but they can easily lead to misinterpretation [25-27].

Fourier Transform Infrared Spectroscopy

An alternative technique that can be applied is FT-IR (Fourier transform infrared) spectroscopy. An IR laser beam is send through the sample surface, partially absorbing the IR light. The absorbed light causes molecular vibrations characteristic for the analysed material at very specific frequencies. Upon collection of the light going through the sample and subtracting it from the original beam, a spectrum is formed that is unique for a specific molecule (so-called molecular fingerprint). Peaks at a certain frequency present in the spectra, give more information on the different functional groups present. When used for plasma technology, if applied in the right way, it can give more information about new functional groups that are introduced via plasma treatment and expose the modifications that have been done to the functional groups already present. The downside lies in the fact that the analysis depth (500-1000 nm) is an order of magnitude higher, thus resulting in a low, often insufficient sound to noise ratio, making it hard to perform a reliable analysis.

Atomic Force Microscopy

Besides the chemical composition, also the surface morphology plays a crucial role in cell adhesion and proliferation processes. Changes in roughness as small as 5 nm can make the difference between success and failure of a tissue engineered system. There are several methods available for visualisation and quantification of the surface morphology. One of the most popular methods for quantifying the surface roughness and generating visual images up to the nm level is atomic force microscopy (AFM).

AFM measurements are based on a laser beam that is reflected on an oscillating cantilever onto a detector. As the tip of the cantilever goes over the sample surface, it will be subjected to changes in interatomic forces. These forces bend the cantilever, giving cause to a shift in the reflection spot of the laser beam. By linking the shift of the reflection spot to a change in sample surface height, a very accurate mapping of the surface is possible (see figure 3). Next to imaging the technique can also be used for quantifying the roughness and measuring attractive and repulsive forces.

Secondary Electron Microscopy

One of the most popular visualisation techniques used in the biomedical industry is secondary electron microscopy (SEM). A beam of electrons is focused via a number of mirrors onto the substrate. The impact of the electrons on the sample surface causes the emission of secondary electrons which are collected onto a detector. These secondary electrons can form high resolution images up to the nm-scale. Not only is it used for mapping the surface morphology, it is also an important technique for cell characterization.

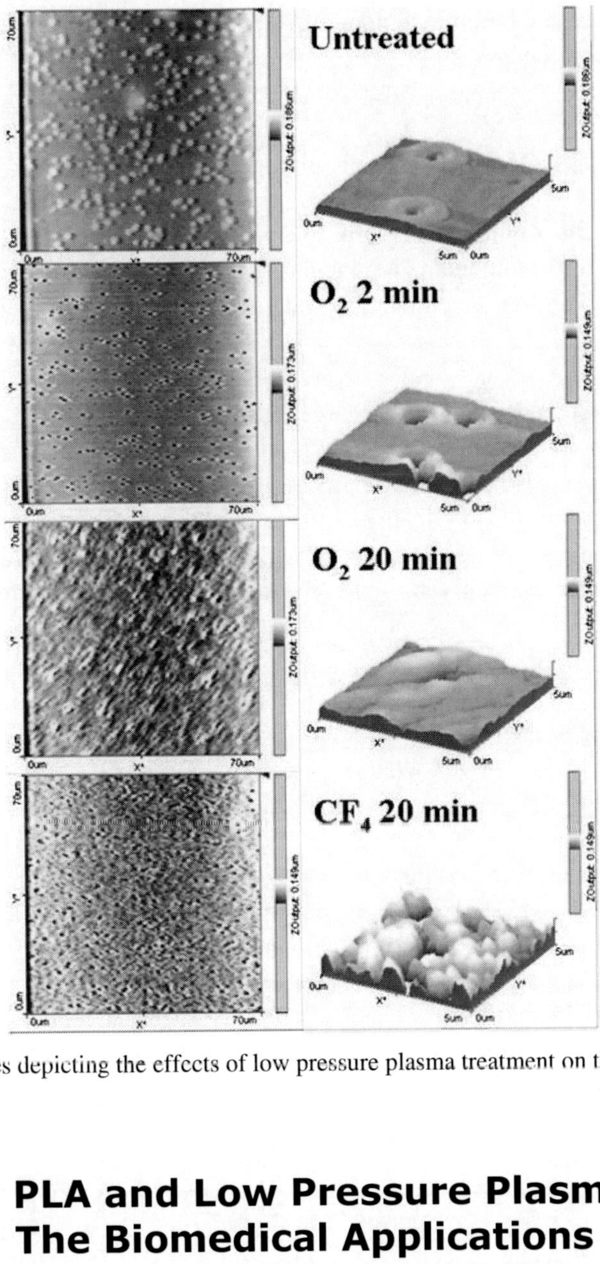

Figure 3. AFM images depicting the effects of low pressure plasma treatment on the surface topography [88].

3. PLA and Low Pressure Plasma: The Biomedical Applications

When combing through the literature for papers on the combination of PLA and plasma, a wide variety of biomedical applications has been studied by labs all over the world. Numerous reactor designs, based on the sources described in the above sections, are used for plasma treatment in order to benefit from one or more of the plasma induced effects on the PLA surface. In order to keep an overview on all the different applications, the chapter will start with those publications focusing solely on the physical changes plasma induces on the PLA surface. This will be followed by the papers dealing with the direct effect of the plasma on the PLA's histological properties. Finally there will be an overview on the use of plasma as a grafting and polymerization technique or to covalently bond specific (macro)molecules on PLA substrates for biomedical purposes.

3.1. Low Pressure Oxygen Plasma

Oxygen is the most popular feed gas for low pressure treatment of PLA, followed directly by argon and anhydrous ammonia. The popularity of these three discharge gasses has been stimulated by the availability of commercial systems working solely with these 3 [28-41].

Wettability

Most papers using an oxygen discharge have performed a surface characterization, starting with a static water contact angle (WCA) study. For non-treated PLA most studies agree on a WCA of about 80°. During treatment, the contact angle decreases progressively to values between 10° and 50° for treatment times between 2-20 min and discharge powers between 10-90 W, which is quite a broad region [20, 21, 28-30, 36, 37, 42-44]. Ferreira et al., mention this difference explicitly by stating that for histological studies, the goal is not looking for the lowest contact angle, or highest roughness, but for the parameters resulting in the best cell response [30].

Surface Morphology

The WCA study is usually followed by a visualisation of the surface, either with AFM, SEM or both. Oxygen plasma in general are well-known for their excellent surface etching properties, introducing a sub-micron roughness which is considered to boost cell growth and proliferation. Mattioli et al., describe the changes in surface morphology as the apparent formation of nano-pillars [21]. Most other papers simply state that there is an increase in surface roughness upon visual analysis of the recorded images. According to Riccardi et al, the plasma will preferentially sputter the amorphous regions of the polymer film, thus resulting in a distinct increase in roughness [45]. When treated for longer periods of time, again a decrease in surface roughness is noted, due to the eventual etching of the crystalline regions [30] (See figure 3).

Chemical Composition

The physical characterization, using the WCA, AFM and SEM results, is linked to a chemical surface analysis (see section 2.3.) involving XPS and IR measurements in only a few papers. For the untreated samples a theoretical O/C ratio of 0.5 would be expected, which is experimentally confirmed by Zhao et al., [29]. After treatment, all studies agree on an increase in oxygen content on the surface, but there is no consensus on the amount of oxygen incorporated, nor as to what kind of function groups and to what extent they are incorporated [20, 28-30, 42, 44].

In-Vitro

When browsing the literature on PLA and non-thermal plasma, it becomes evident that the research is mainly focused on the histological impact of the treatment, rather than on the treatment itself. For the analysed studies using oxygen as a feeding gas, 6 different cell types were seeded onto the samples for cell adhesion and proliferation essays: MG63, FRC, CHO, B65, M3T3, hMSC… [20, 28, 36, 38, 40-42, 44, 46].

Although it could be interesting to discuss each article, it would lead to far and the conclusions for all of them are more or less the same. The adhesion essays show that in the first 24 hours, the cell density for the plasma treated samples lies significantly higher for the plasma activated PLA compared to the untreated samples. The study using CHO, performed by Yamaguchi et al, shows that the plasma activation results in similar cell adhesion compared to the standard tissue culture PS dishes [44]. After the first 24 hours and up to 7 days, the proliferation rate is more or less the same for all cells tested, with the exception of the hMSC cells, which show an overall superior proliferation rate.

When comparing the cell morphology after 7 days of incubation, there is a major difference: the cells on the untreated samples are small, round and no network is formed between different cells. For the cells proliferated on activated samples, cells flatten out and form an intercellular web. One of the direct consequences of this difference in cell morphology has been studied by Wan et al., they exposed the PLA, seeded with M3T3, to shear stress. For the untreated polymer films this resulted in extensive detachment of the cells form the surface, while for the treated PLA the cells extrusion was retracted to the body and al changes happened very slowly, indicating a more stable cell culture [36].

Overall it can be concluded that the changes in surface- chemistry, wettability and roughness induced by the oxygen plasma, result in benign histological properties.

3.2. Low Pressure Ammonia Plasma

The big difference between oxygen and ammonia plasma is the introduction of nitrogen containing species into the plasma and a possible incorporation of amines onto the PLA surface, which are considered beneficial for cell adhesion and proliferation.

Wettability

In contrast to the WCA results obtained from the oxygen treated PLA, there is more of a consensus on the contact angles for the ammonia treatments. Using an RF discharge, WCA are found starting from 60° up to 20° [28, 32, 36, 47-49]. In contrast to other gasses, prolonged exposure to the NH_3 plasma does not lead to a fixed angle, but a significant re-increase in WCA is noted. The lowest contact angle is obtained at a power of 50 W and an exposure of 2 min in all studies. Jiao et al., contribute these higher angles at higher energy densities to the thermal degradation of PLA, causing a decrease in the incorporation of polar groups [47]. Compared to the oxygen plasma, the contact angles are slightly higher, but as mentioned before, it is not always the highest density in polar groups that lead to the best histological results.

Surface Morphology

The literature describing the changes in surface morphology is limited, but those papers that have a surface morphology study included come to the same conclusion as the oxygen plasma: there is an increase surface roughness and a fine grain structure is introduced [48]. More interesting is the SEM study of Wan et al., that showed a visible degradation of a PLA scaffold when exposed to powers higher than 20W [36]. This was confirmed by the AFM study of Jiao et al., (50 W-2min) that showed the first signs of film destruction and the softening of the borders between crystals [47]. Linking the changes in surface topography with the wettability studies, confirms that a careful selection of discharge power and treatment time is advised.

Chemical Composition

As mentioned in the beginning of this section, ammonia plasma are mainly used to incorporate nitrogen functionalities into the PLA surface. The focus of the chemical surface composition therefore lies onto the analysis of the nitrogen peaks. In this case no literature was found using IR as an analysis technique, so all results presented are measured via XPS [28, 32, 33, 47-50].

In general, the effectiveness of the nitrogen incorporation can be linked again to the changes in surface morphology and WCA: an overexposure (Power and time) of the PLA leads to a reduction of nitrogen functionalities compared to more moderate operating conditions. Different papers report an incorporation of 5-10% of nitrogen associated with a decrease of both oxygen and carbon. The deconvolution of the nitrogen peak at 400 eV learns that the majority of the nitrogen is incorporated as primary amines (75%) and equal amounts of imines and amides (10-15%) and this for a power of 50W and a treatment time of 2 min(See figure 4) [33]. As mentioned before, an increase in power leads to a decrease in nitrogen content and causes a shift from primary amines to amides.

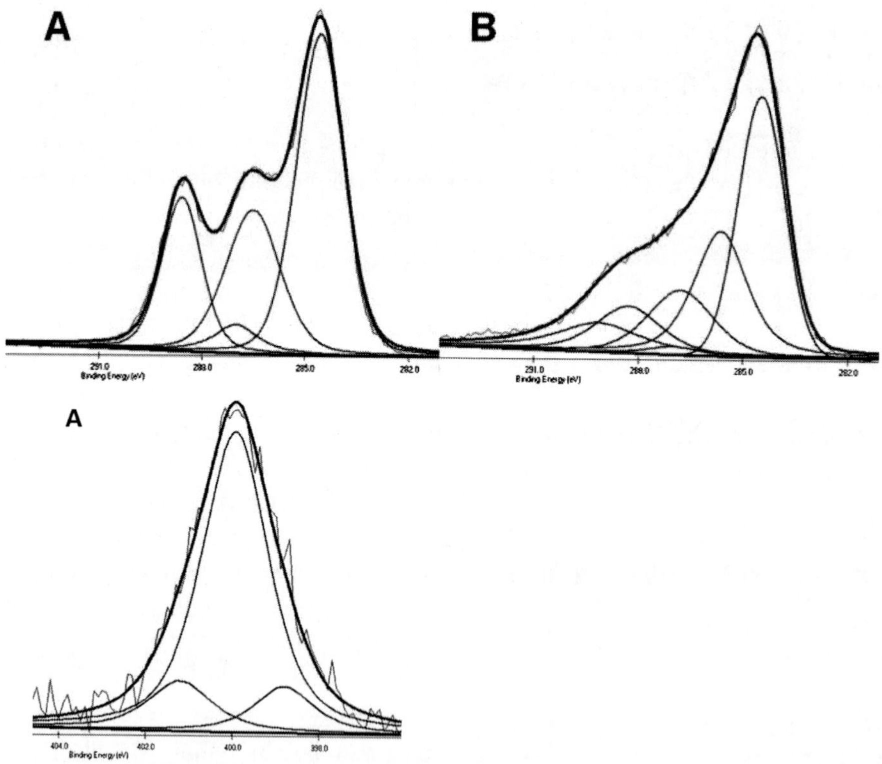

Figure 4. Deconvolution of the C1s and N1s XPS peaks, proving the presence of Amine and Amide functional groups [89].

Jiao et al., performed a chemical reaction using an acidic orange 7 solution to form a complex between the primary amines and the acid dye [47]. This was followed by washing the samples with a NaOH solution, desorbing the dye. The NaOH solution is exposed to light with a wavelength of 492 nm and an absorbance spectrum was recorded. Based on this procedure an amino group density on the PLA surface of 1.4 10^{-9} mol/cm^2 was found. Combined with the XPS analyses it is proven that ammonia plasma are able to incorporate primary amines onto a PLA surface.

Biomedical Applications

In contrast to the oxygen treatments, not only histological studies were performed, but also protein and peptide adhesion was investigated.

The histological studies show a similar enhancement in the adhesion essays compared to the oxygen plasma for all cell types tested: M3T3, HUVEC, rbMVEC, BAEC, BSMC, fibronectin, H1444 and rat calvarial cells [28, 31, 32, 47, 49-52]. The proliferation essays on the other hand differ compared to the oxygen proliferation essays in such a way that the ammonia plasma also significantly enhances the proliferation process, with the exception of the M3T3 cells, up to and beyond the growth rate of tissue culture PS dishes.

Several groups went beyond the 2D-films and started investigating cell properties on 3D-scaffolds [31, 32, 50]. Not only were they interested in adhesion and proliferation essays on the scaffold surfaces, but even more so in the migration efficiency of the cells into the scaffolds. Cheng et al., performed a very thorough study of the cell infiltration into the scaffold, both in-vitro and in-vivo [50]. 20 μm thick slices were cut from the scaffolds and cells were coloured with a fluorescent dye. In figure 5, 3 different cross sections are included. The untreated one at the top shows a very low density of cells present and almost no infiltration of the cells. The cross section in the middle, showing the effects of an Ar treatment, depicts a much higher cell density and reasonable infiltration. The lower one, being the result from the ammonia treatment, shows both excellent cell proliferation and infiltration. The same study was performed in vivo on rats and the same results were obtained, though with a lower efficiency. Therefore they concluded that an excellent infiltration in-vitro is required in order to have success in-vivo and that the combination might lie in a combined use of plasma treatment and the incorporation of growth factors into the scaffold.

The protein and peptide adsorption studies found give mixed result on the plasma treatment of PLA. Xu et al., investigated the anchorage of GRGDS peptides onto PLA scaffolds before and after an ammonia plasma treatment. After the treatment they saw an excellent conjugation between the peptide and the polymer surface, caused by the amide bonding between the incorporated primary amines and the carboxyl functionalities present in the peptide [33].

Sarapirom et al., did some research on the adsorption of the HSA protein before and after treatment and came to the conclusion that the protein prefers non-polar, untreated, surfaces. This result should be stressed, because it points out that not every treatment is successful for every application, emphasizing the need for application tailored research [48].

Finally Yang et al., used an ammonia plasma treatment for the immobilisation of collagen onto a PLA surface. Compared to a collagen coating, the M3T3 cell proliferation showed no significant differences, but once the samples were rinsed in PBS solution for some minutes, some interesting changes were noted. Before and after rinsing, the cells number on immobilized collagen did not differ significantly. For the normal collagen coating on the other hand, most of the cells were washed away. This proves that plasma technology is able to anchor collagen tightly onto the surface, making the treated scaffold a better match for tissue engineering purposes.

Overall it can be concluded that ammonia plasma treatments are able to incorporate primary amines onto a PLA surface, making it a more versatile treatment and resulting in a better proliferation compared to oxygen treatments. Although it leads to positive effects in most cases, a case to case study is required to guarantee its success.

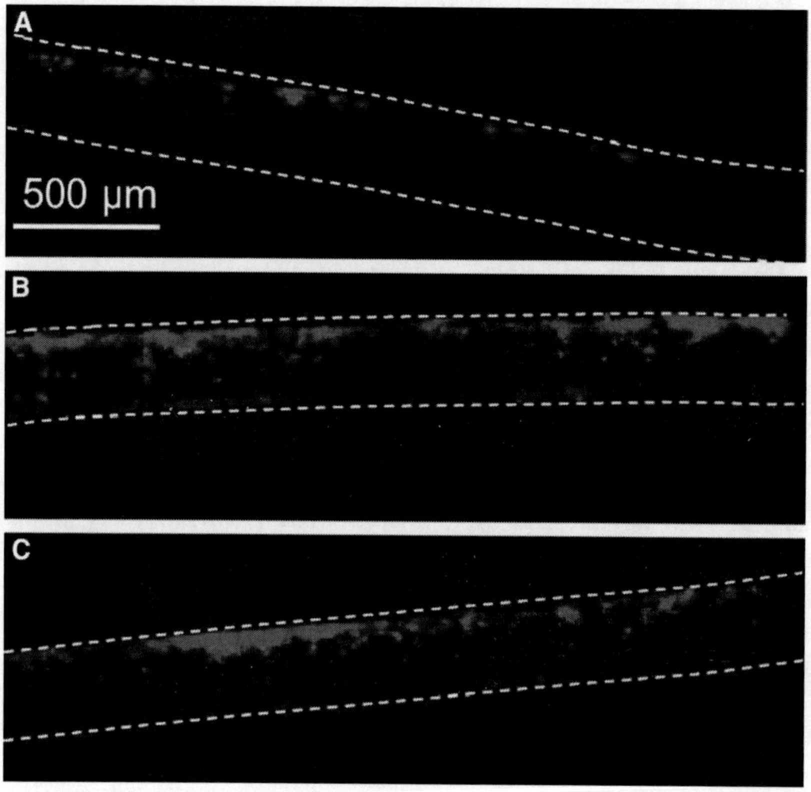

Figure 5. Different fluorescent cell essay cross-sections of: A) untreated scaffold, B) Ar plasma treated scaffold and C) ammonia plasma treated scaffold [90].

3.3. Low Pressure Argon Plasma

Of the 3 most used low pressure plasmas, Ar plasma is considered to be the mildest treatment technique. Via a cascade of secondary processes, reactive sites are introduced onto the surface that, once exposed to ambient air, result in the incorporation of a whole set of oxygen containing functional groups [53].

Wettability

The more mild conditions of the Ar plasma treatment are clearly reflected in the WCA results. Whereas oxygen and ammonia plasma reached contact angles below 20°, the Ar treatment results have a saturation contact angle between 25° and 50°, depending on the discharge system used [29, 35, 50, 53-56]. Rather than differences in surface topography, this difference of 30°, in some cases, is most likely caused by the differences in the kind of functionalities that are incorporated, which will be discussed in one of the following paragraphs.

Surface Morphology

The changes in surface morphology that are introduced by the argon plasma are comparable to the oxygen and ammonia treatments. Different papers describe the phenomenon of different etching rates between the crystalline and amorphous regions of the film resulting in the exposure of the crystallites present in the film [29, 35, 54, 56]. Zhao et al., did a comparative study between Ar and oxygen plasma treatments and for the surface morphology they found that the etching of the PLA is less pronounced compared to the oxygen etching [29]. Slepicka et al., found an increase in roughness from 6.9 nm to 11 nm, which was in line with the expectations. More surprisingly, a gravimetrical analysis of the treated samples showed that, using a DC discharge of 10 W, the etching is capable of removing up to 75 nm of material over a period of 4 minutes. This shows that the top layers are etched away completely and that the underlying layer is etched at different rates, depending on the crystallinity, as stated before [56].

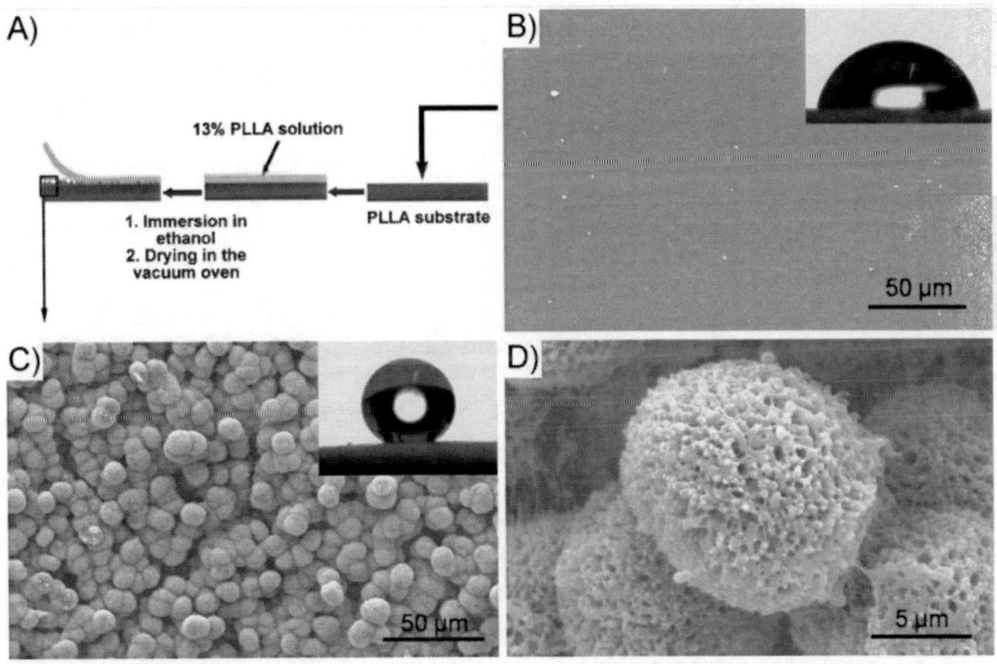

Figure 6. A) Schematic representation of the experimental process to produce superhydrophobic surfaces. B) SEM image of the smooth PLLA surface. C) SEM image of the rough surface. D) Magnified SEM image of a protrusion on the superhydrophobic surface. Insets: Water drops on the corresponding surfaces [91].

In the next paragraph, special attention will be given to an article published by Song et al., In their study they synthesized super-hydrophobic PLA films using a novel polymer solution-air casting method (See figure 6) [57]. In their study, they investigated the effect of argon plasma on the wettability properties of the super-hydrophobic film. Starting from a WCA above 150° the plasma treatment results in a progressive decrease of the contact angle up to and under 5°. Compared to the treatments on smooth films, the effect of the Ar plasma treatment is greatly enhanced. The reason for mentioning this research here, is to stress on the combined effect surface roughness and plasma treatments have on the wettability of a PLA surface. In the application section, more attention will be given to the histological results found in this study.

Chemical Composition

As mentioned in the introduction, the formation of radical sites onto the sample surface occurs through a cascade of secondary reactions as can be seen in figure 7. Once exposed to the ambient air, oxygen containing functional groups can be formed at the reactive sites.

When browsing through the literature dealing on the chemical surface analysis of Ar plasma treated PLA, it becomes apparent that most studies do not result into the incorporation of an extra amount of oxygen containing functionalities, nor are they able to introduce new functionalities [34, 35, 53, 54, 56, 58]. Different groups even note a decrease in oxygen content, suggesting that degradation of the PLA takes place. Linking this to the wettability and surface morphology analyses, it would suggest that the Ar plasma etching and the resulting increase in surface roughness are the main causes of the decrease in WCA.

Those treatments that were able to increase the oxygen incorporation, find similar results as the oxygen treatment, with the biggest increase of oxygen being contributed to carboxyl functional groups [29, 55].

In general, the surface analysis study clearly shows that treatment parameters have to be carefully selected in order to avoid degradation of the polymer material and to obtain significant changes in chemical composition.

Biomedical Applications

The articles dealing on the histological effects and adhesion of peptides are limited and just as the surface analysis, the results are somewhat mixed. Slepicka et al., seeded VSMC cells onto the Ar plasma treated surface, and although the first 24 hours suggest a better adhesion, after 7 days no differences were found between treated and untreated samples, indicating that PLA is an excellent material to grow smooth muscle cells to begin with [54, 56]. Cheng et al., whom have been discussed before for their excellent study on the cell adhesion and proliferation of BAEC and BSMC cells on ammonia plasma treated, also used Ar plasma for treating their PLA scaffolds and their comparison between treatments showed that on the scaffold surface Ar plasma would result in a significantly better proliferation. For the cell infiltration, as depicted in figure 5, excellent cell infiltration was found, similar to the NH_3 treatment and superior compared to the untreated PLA. In the in-vivo studies, the differences between the in-vitro treatments were no longer detected, showing again the importance of in-vivo testing [50].

Ding et al., used the Ar plasma treatment to successfully immobilize chitosan onto the PLLA surface. The cell tests, using L929 and L02 cell lines, gave no significant difference in proliferation rate compared to the cells seeded onto glass plates. Also the cell morphology was round shaped, indicating the lack of cell differentiation [55].

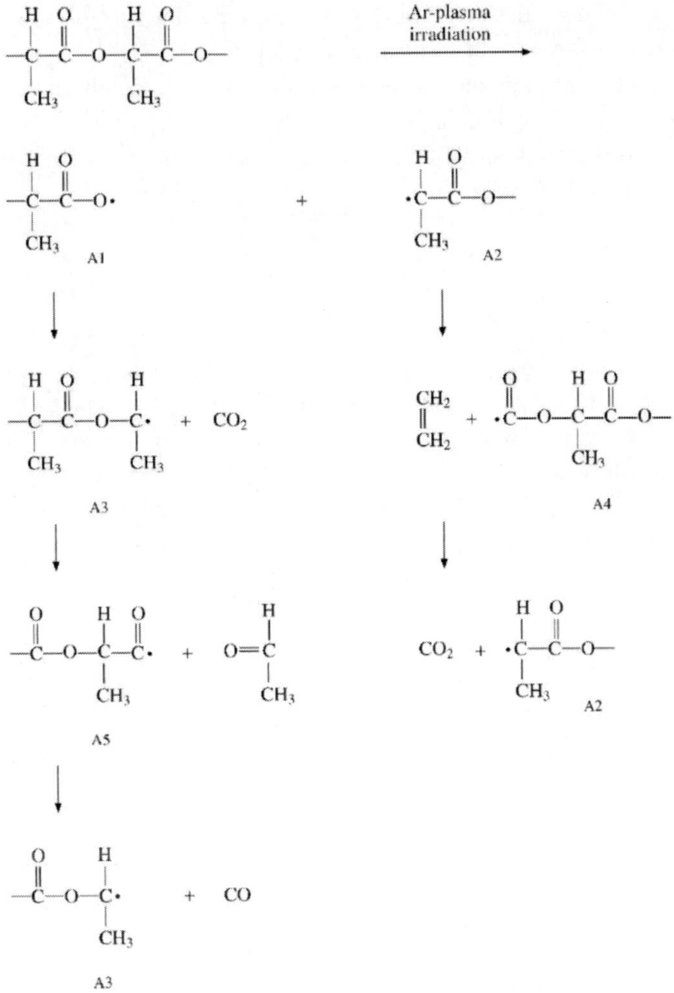

Figure 7. Possible reactions occurring on the PLA film surface on Ar-plasma irradiation [92].

Ho et al., used plasma treatment as a tool for the immobilization of RGDS and RGES peptides. They compared the adhesion, growth and proliferation of ROS cells onto untreated, plasma treated and peptide immobilized surfaces. After the first 24 hours, the surfaces that were having RGDS anchored onto them resulted in the highest cell density. The RGES and plasma treated samples gave similar results, indicating that the RGES had no significant influence and that the changes in surface chemistry and morphology induced by the plasma were the main contributors to the enhanced growth compared to the untreated samples. After 4 weeks the same trends were observed and after calculating the growth rate curves they concluded that RGDS coated surfaces not only lead to a better adhesion but also to a faster cell growth and division [59].

As mentioned in the surface morphology section, Song et al., also performed cell adhesion and proliferation tests, using a L929 cell line, on their super-hydrophobic and hydrophilic PLA surfaces. Results show that on the hydrophobic surfaces cell growth is poor and the cell remain small and round. On the Ar plasma treated surface, a significant increase of cell growth is noted and the cells exhibit a more flattened and extended morphology [57].

Overall, this research shows that combining surface structuring with plasma treatment allows tuning of biomaterial surfaces to their specific application.

It can be concluded that Ar plasma treatment can be a viable alternative, if the reaction conditions have to be mild and if changes in surface morphology play a more important role than differences in surface chemical composition. Again one has to be careful for sample degradation and what kind of biomedical application it is used for.

3.4. Hydrophobic Surfaces

As mentioned multiple times in the previous parts of the chapter, the increase of surface hydrophilicity does not always result in better cell growth and adhesion. In other cases, the adhesion of cells (crf heart valves, insides of stents, dialysis equipment…) has to be avoided at all cost, as it would lead to a reduced functionality or complete failure of the material application. The method of choice to generate (super)hydrophobic surfaces is by using a fluorinated gasses to form the discharge. The most popular gas would be CF_4, but also other gasses such as SF_6 have been used to treat PLA. It is even possible to generate super hydrophobic coatings using non-fluorinated gasses, but articles, using these techniques on PLA, have yet to be published.

Wettability

Treating the surfaces with CF_4 or SF_6 plasma both have the same effect on the wettability of PLA. An increase between 20° and 40° in WCA up to values of 115° were found, depending on the treatment time and power used [21, 60-62]. As fluorinated species are known for their extensive etching, their effect on the surface morphology and thus the indirect effect on the wettability should not be neglected and will be discussed in the next section.

Surface Morphology

Of all the articles describing the changes in surface morphology, the publication of Mattioli et al., was the most clarifying, as it compares the topography after oxygen and CF_4 plasma treatments [21]. In the chapter part on low pressure oxygen plasma, the technique is praised for its excellent etching properties. Putting the pictures underneath one another, as can be found in figure 2, it clearly shows that the etching effect is more profound for the CF_4 treatment. The same increase in surface roughness can be found in the other articles [60-62].

Chemical Composition

The main difference between fluorinated gas plasma treatments and other plasma treatments is self-evidently the incorporation of fluorine onto the PLA surface. XPS measurements show an increase up to 25% of fluorine. Whereas the elemental composition shows no real changes in carbon concentration, it is mainly the oxygen that is being replaced. Chaiwong et al., performed ATR-FTIR spectroscopy after treatment with SF_6 plasma and found C-F stretch vibrations, indicating that the fluorine is covalently bond onto the PLA surface. Ageing studies show that after 4 weeks, the amount of fluorine present on the surface does not change significantly, supporting the statement that a stable bond is formed between fluorine and the PLA surface [60].

Biomedical Applications

Although the fluorination technique has been extensively used on other biomaterials [37], only 1 article was found describing the antibacterial properties of the fluorinated coatings. Boonyawan et al., tried to grow E. Coli bacteria onto treated PLA surfaces. After fluorination they saw a decrease in bacterial adhesion of almost 40%, while an antibacterial efficiency of more than 92% was achieved [62].

Although the haemocompatibility of fluorine coatings has been studied on other substrates, the biocompatibility of CF_4 plasma treated PLA has yet to be tested in-vitro and in-vivo [63]. The first results indicate that these kinds of coatings can be used for antibacterial purposes.

3.5. Other Low Pressure Discharges

Next to the conventional discharge gasses, research groups have been experimenting with a number of other gasses and tested the histological effects. In what follows, each gas will be discussed briefly and will be linked to the discharge gas it best resembles.

Low Pressure Air Plasma

Air plasma, being a mixture of oxygen and nitrogen, can be considered as a solid alternative for the oxygen plasma treatment. Even though the discharge gas contains around 80 % of nitrogen, very little nitrogen functionalities will be incorporated, as oxygen is the far more reactive species and dominates the radical formation and functional group incorporation processes.

When comparing the changes in wettability between air and oxygen plasma treatments, similar values, lying between 10° and 15°, are found [64]. No AFM measurements were found, so no surface morphology information can be given. XPS measurements were performed and a slight decrease in oxygen content was noted, that was compensated by the incorporation of 3-5% of nitrogen containing functionalities. As the number of polar groups did not change significantly, Demina et al., contributed the changes in wettability and surface energy to an increase in surface charge density. Upon performing a histological study, using mouse fibroblasts (L929), they found that the cells grew and differentiated better after treatment [64].

Low Pressure CO_2 Plasma

Another gas that is commonly available and is considered waste in most industrial processes is CO_2. 2 papers were found using CO_2 as a discharge gas, of which only 1 focussed on biomedical applications [61, 65]. After treatment, the contact angle decreased from 84° to 48°, similar to Ar plasma treatment. FTIR spectroscopy resulted in a completely different fingerprint, indicating that CO_2 is incorporated onto the surface. The characteristic ester peak at 1730 cm^{-1} increased dramatically in intensity and together with a more subtle appearance of the OH band at 3300 cm^{-1} this spectrum suggests an increase of different oxygen containing functionalities, similar to air and oxygen plasma treatments. For their histological studies Khorasani et al., used 2 different cell-lines, B65 and L929, and examined their adhesion and proliferation. For the B65 cell line, better adhesion and proliferation was noted, cells were flattened out and an extensive network was found. For the L929 line, no significant differences were found compared to the untreated PLA.

If an overview is made of the different treatments were the L929 cell line is used, it becomes clear that different discharge gasses and parameters result in extreme differences concerning cell adhesion, proliferation and morphology, making it not always straightforward as to what treatment should be considered [55, 57, 65].

4. Medium/Atmospheric Pressure Plasma

Atmospheric pressure plasmas were already used by Siemens in the second half of the 19th century for the ozone generation, but only in the last 25 years there has been a renewed interest in using the technique for material surface modification, as an alternative to low-pressure plasma treatment systems. The lack of extensive vacuum equipment makes it time and cost-effective and allows easy incorporation in production lines.

As mentioned in the general introduction, plasma modification at elevated pressure has proven its effectiveness on aliphatic and aromatic polymers of all kind and together with the growing interest in biodegradable materials, there have been a number of publications using atmospheric pressure plasma treatments on polylactic acid of whom a selection will be discussed in the following chapter section.

Due to the limited amount of studies available in literature dealing on the plasma treatment of PLA at atmospheric pressure conditions, a different subdivision will be applied. The first part of this chapter section will give an overview of parallel plate systems used at atmospheric pressure. In the second part, special attention will be given to a special set-up that has gained a lot of interest in the field of plasma technology: the plasma jet.

4.1. Parallel Plate Systems

Most systems operating at atmospheric pressure conditions are based on the DBD principle (see section 2.2). The most basic set-up available consists of 2 planar electrodes, of which 1 or 2 are covered with a dielectric material. This system can be operated in a closed reactor, giving control over the atmosphere present, or in ambient air, making it more easily applicable.

The discharge gasses used for treatment are similar to the low pressure systems, with the exception of helium. Helium is quite popular, as it allows for very stable glow-like plasma. The downsides of helium as a discharge gas, is that it is less effective in introducing polar functionalities as the surface and due to impurities present, often results in a mediocre reproducibility.

Before starting with the analysis of results, it is important to make the distinction between atmospheric pressure and medium pressure treatment. Medium pressure treatments, operating at a few kPa, do not need the extensive vacuum equipment used for low pressure systems (a rotary vane or membrane pump is still required), while at the same time more stable plasma are obtained compared to atmospheric plasma treatments. Due to the presence of a pumping procedure, the treatment is not as time efficient as its atmospheric pressure counterpart [66].

Morent et al., performed a comparative surface characterization study on plasma treated PLA at medium pressure (5kPa), using He, Ar, air and N_2 as discharge gasses [67, 68]. The WCA study shows a progressive decrease in contact angle as a function of energy density. Big differences are found between gasses: 59° and 60° for air and argon treatments, 35° and 31° for helium and nitrogen treatments. AFM measurements showed small differences between treatments, therefore these WCA values primarily have to originate from differences in surface chemical composition. XPS measurements showed that the decrease in WCA, due to air and argon plasma treatments, was caused by the incorporation of extra oxygen containing functionalities (+ 5-7% O) in the films. For the helium and nitrogen treatments no significant increase in oxygen was noted, but for both treatments the incorporation of nitrogen functionalities was detected (+3-6% N), hence the difference in WCA. Similar results were found by Hergelova et al., for treatment in ambient air (55°-60°), by Vergne et al., for treatment in nitrogen (27+/-3°), both at atmospheric pressure [69-71].

Comparing these results with the low pressure plasma treatments one could conclude, based on the WCA, that a higher pressure results in a reduced treatment efficiency. The XPS results on the other hand indicate that the incorporation of functional groups is at least as favourable. In this case in-vitro or in-vivo studies would be able to make a distinction in efficiency between treatments, but for the parallel-plate set-up, only 2 histological studies have been performed. Jacobs et al., used HFF-1, human foreskin fibroblasts, cells, seeded on PLA plasma treated films, using 3 different discharge gasses (Ar, He and air) [22]. The study finds very similar results compared to low-pressure treatments, showing a better cell adhesion in the first 24 hours. After 7 days, the proliferation study shows no significant differences anymore between untreated and treated PLA. Phase contrast and fluorescent microscopy indicate superior morphology after 24 hours, independent of discharge gas. After 7 days both untreated and treated samples show the same morphological features. Nakagawa et al., seeded MC3T3 cells on air and CO_2 plasma treated samples and found similar results compared to similar studies performed at low pressure [72].

Two histological studies are insufficient to draw any major conclusions. Therefore, the chapter will continue with an overview of that other, maybe even more popular set-up: the plasma jet.

4.2. Atmospheric Pressure Plasma Jet

The biggest advantage of the jet compared to all other treatments is its mobility, making it possible to homogeneously treat complex 3D structures. Furthermore, there is no pumping procedure, making it cost and time effective.

As just mentioned, plasma jets are often used for treatment of 3D structures. In this perspective, the electrospinning of scaffolds is by far the most popular application in the biomedical field. To analyse these materials, contact angle goniometry cannot be used in the same way, making it difficult to make an objective comparison on the wettability. Other techniques (AFM, SEM, XPS etc.) do have the needed spatial resolution to give an adequate analysis of the changes caused by APPJ treatment.

AFM analysis of the surface morphology results in images that are identical to the ones found for other plasma treatment techniques (see section 3). By changing the movement speed of the jet, the surface morphology can be tailored from completely flat up to the same topography found for fully saturated samples [73, 74].

XPS analysis reveals that plasma jets have fewer problems with the incorporation of oxygen into the sample, up to a point were O/C ratios higher than 1 are found. This indicates that the presence of oxygen in the surrounding atmosphere can have a beneficial influence on the incorporation of polar functional groups [73-77].

The in-vitro studies again indicate results that are similar to other treatment techniques: the seeding of MC3T3 fibroblasts leads to a better adhesion in the first 24 hours, and a similar proliferation rate after 7 days with enhanced morphological properties [73]. Other groups used the plasma to generate anchor sites for chitosan/gelatine. Both in-vitro and in-vivo studies show excellent stability of the immobilized proteins and superior cell adhesion (C2C12 and nerve regeneration respectively) and proliferation compared to untreated and uncoated samples [74, 76].

In general it can be concluded that medium and atmospheric pressure systems form a valid alternative for low-pressure treatments, giving similar results concerning wettability, polar group incorporation, morphology modification and histological response.

5. Plasma Coating

The main principles behind plasma grafting and polymerization are already explained in part 2.3.

One of the major disadvantages of non-thermal plasma treatments is the limited stability of the treatment. Over time, the surface will try to restore its initial surface energy, resulting in a decreased wettability and the migration of polar groups. To overcome this instability, research groups have turned their attention to using the formed reactive sites (grafting) and the plasma itself (polymerization) as initiation sites for polymerisation reactions.

5.1. Plasma Grafting

In parts 3 and 4, the use of plasma treatments for immobilization of polysaccharides and proteins such as chitosan, cellulose and RGD have been mentioned. Some authors would also consider this a form of plasma induced grafting, being limited to a 1 step reaction. In what follows next, an overview will be given on the use of plasma technology for the grafting and polymerization onto polylactic acid substrates.

Single-Step Grafting

Some research on the immobilization of macromolecules has been discussed before and will not be gone through again in detail, but as most of them analyse the histological effects both on plasma treated and plasma grafted samples, some general trend can be found [28, 29, 55, 78]. In every single case the grafted substrates are equal or superior in performance compared to plasma treated samples. This is no real surprise, as the macromolecular structure of proteins and sugars is well-liked by cells and stimulate adhesion, proliferation and differentiation. Therefore they have been the choice of material or coating for many tissue engineering labs.

In this next paragraph there is a special mention for the work of Yang et al., [28]. In their research, a comparative histological study was done between PLA samples that were coated with collagen, with or without ammonium plasma pre-treatment. 7 days after the seeding of the MC3T3 cells, no real differences in proliferation were found. After rinsing with a PBS solution, the plasma grafted samples showed no significant difference in cell count. The samples that had been immersed in a collagen solution without pre-treatment gave cause to a significant decrease in cell count after rinsing. This research proves that the plasma pre-treatment results in a tight bonding between the collagen and the PLA, therefore guaranteeing an optimal functionality of the biomaterial in a more active environment.

Traditional Grafting

For plasma grafting in the traditional sense, acrylic acid and allylamine are two popular precursors, as the functional end-groups of the resulting polymers, carboxylic acid and primary amine respectively, are well liked by a great variety of cells and can be used as intermediates for follow-up reactions [78-81].

Barry et al., performed both plasma grafting and plasma polymerization using allylamine as a precursor to deposit polyallylamine on porous PLA scaffolds and compared the distribution of the coating as well as the histological performance [81]. In what follows here, only the grafting part will be discussed, while in the section on plasma polymerization, a comparison will be made between the different approaches. The XPS analysis shows a homogeneous incorporation of nitrogen al throughout the scaffold, though at concentrations of only 2 percent, indicating a low grafting efficiency (see figure 8).

Figure 8. Nitrogen concentration at the surface as determined by XPS at set intervals across the internal diameter of grafted and ppAAm-deposited scaffolds (20 W unwashed). These were calculated at set intervals along the internal diameter (shown as a dotted line in the computed tomography image of the PLA scaffold, following mechanical sectioning. The photoelectrons were collected from a 0.3 x 0.7 mm rectangle oriented with the short side parallel to the direction of translation [93].

Deconvolution of the peaks shows that the amine functionality is preserved, even after rinsing with water. The cell essay shows that 24 hours after seeding, there is an increase of about 40% in cell adhesion compared to the untreated samples. SEM imaging showed a mixed cell morphology, both round and flat, and no visible infiltration of the cells into the inner parts of the scaffold.

Hydroxylapatite

Another interesting application that has not yet been discussed in this chapter is the enhanced growth of apatite on PLA. (Hydroxyl)apatite is a mineral consisting out of calcium phosphate and plays a central role in tissue engineering for implant applications as it stimulates osteointegration [82]. Yokoyama et al., used an oxygen plasma to introduce extra functional groups onto the PLA surface, after which the samples were alternatingly dipped in Ca^{2+} and $(HPO_4)^{2-}$ ion solutions [79]. Due to electrostatic forces, the ions were adsorbed. After the dipping process, the samples were stored in a simulated body fluid (SBF) solution. After 24 hours, analysis showed the growth of a dense apatite layer onto the PLA surface and this in contrast to the samples that were not exposed to plasma. Adhesion tests showed that the grown apatite layer adhered better to the PLA surface compared to commercially available apatite (HAPEX™). Park et al., used a different approach where the PLA was submitted to acrylic acid plasma grafting, followed immediately by immersion in SBF solution. After different time periods (5-10-15 days) due to the interaction between the COOH functional end groups of the grafted polyacrylic acid and the ions in the SBF solution a stimulated growth of apatite was found compared to the untreated PLA. In-vitro studies showed a superior cell proliferation at all time for fibroblasts, osteoblasts and chondrocytes. These two different approaches on the growth of an apatite layer on PLA shows again the versatility of plasma technology techniques.

In general it can be concluded that for flat structures, plasma grafting is an excellent technique to boost the histological performance of PLA and gives better results for in-vitro studies compared to plasma treated samples. For more complex surfaces, such as scaffolds, grafting has a beneficial influence on cell adhesion and although the grafting process results in a homogeneous distribution of functional groups throughout the scaffold, the grafting efficiency is mediocre.

5.2. Plasma Polymerization

The introduction of thin films on biomaterials is something that has been done for a very long time. The typical problems for biomedical coatings are threefold and often counteract. First of all, the coating has to adhere sufficiently to the substrate in order to avoid unwanted migration of coating material through the body. Secondly, any changes to the bulk properties of the biomedical material, such as mechanical characteristics and structural integrity, have to be avoided at all costs, in order to guarantee its optimal performance. Thirdly, no solvents or other chemicals which are not approved for biomedical applications can be used during the process. Due to the lack of solvents and the non-invasive character of the plasma, none of the above mentioned points are considered to be a problem during the plasma polymerization process. Compared to plasma treatment, a wider variety of functional groups can be introduced, while at the same time the substrate can be protected from its environment or vice-versa.

The literature is not (yet) abundantly filled with papers dealing on the deposition of thin films on PLA using plasma polymerization for biomedical applications, but in what follows, some interesting examples will be discussed in more detail [71, 81, 83-85].

Barradas et al., deposited two different kind of coatings based on tetramethylsilane (TMS) and 3-aminopropyl-trimethoxysilane (APTMS) precursors and compared there influence amongst other tests to the osteogenic differentiation of MC3T3 cells on untreated and N_2/H_2 plasma treated PLA [71]. For the deposited TMS coatings they found poor cell adhesion and proliferation, even compared to untreated material, which they could directly link to its hydrophobic properties. The cells themselves were heterogeneously distributed over the surface, which in turn seemed to have a positive effect on the osteogenic differentiation. For the APTMS coatings similar results were found compared to plasma treated samples: a better cell adhesion in the first 24-48 hours followed by a proliferation period that does not differ from untreated material. As for the osteogenic differentiation, both the plasma treatment and APTMS coatings had a negative influence on the differentiation. In general they conclude that a clear link between the physicochemical properties of PLA-modified surfaces and the cellular responses of MC3T3-E1cells could not be established but that the surface chemistry appears to play a more important role compared to surface roughness.

As discussed in the plasma grafting paragraphs, Barry et al., also deposited polyallylamine coatings via plasma polymerization using a low-pressure plasma to deposit the coating, applying two different discharge powers (3 and 20W) [81]. Chemical analysis showed that besides the presence of amines, also a considerable amount of amides was present, even more so for the higher discharge power, which is responsible for a more profound activation and degradation of the monomer. After washing, also nitrite and nitrate functionalities were detected, indicating that also for plasma polymerization a reactive surface is formed. Next to the conversion of functional groups, the washing also caused thinning of the deposited films, especially for the 3W conditions. This was contributed to the lesser cross-linking of the polymer compared to higher discharge powers, resulting in a better solubility of the coating. For the grafting, a homogeneous distribution of nitrogen was found all throughout the PLA scaffold, but with a mediocre grafting efficiency. For the plasma coating, a gradient in nitrogen concentration was found throughout the scaffold, as can be seen in figure 8. This phenomenon was explained by the authors as the inability of the monomer to migrate at the same rate through the micropores. The cell essays showed a significantly higher cell count for the plasma coated samples compared to the grafted samples. SEM imaging revealed that the cells had formed sheets and bridged the pores, completely covering the PLA surface, which was not the case for the plasma grafted scaffolds which only showed cell growth in the outer regions of the scaffold. In general they concluded that the higher amount of nitrogen was directly responsible for the enhanced histological properties of the scaffolds.

There are a number of other papers describing the coating of PLA films and substrates with potential for biomedical applications, but as they do not include an include in-vitro or in-vivo studies they will not be discussed in more detail [83, 85]. Furthermore it has to be stressed that plasma polymerization is less substrate dependant compared to plasma grafting and plasma treatments. Therefore results from other (biodegradable) materials can easily be adapted and applied to PLA substrates. An excellent, more general review has been written by

Da Ponte et al., on the subject of coatings deposited via plasma polymerization [84]. Finally one can make the general conclusion that plasma polymerization is a technique that is able to introduce a wide range of functional groups on polylactic acid films and scaffolds with unique properties concerning adhesion and histological performance.

Conclusion

In this chapter an overview has been given on different plasma related techniques that were used to enhance the biomedical performance of polylactic acid. Most importantly to remember about plasma treatment, at all operating pressures is that in general they succeed in improving the histological performance of PLA, but that each case has to be separately approached, as each cell-line prefers different mechanical and chemical surface properties. Plasma grafting and polymerization offer a wider variety of functional groups, while they offer at the same time the formation of a protective barrier between the implant material and its environment. In our professional opinion, we believe that in the years to come there will be a further shift towards atmospheric pressure systems, with special attention for the plasma jet configurations. As tissue engineering is rapidly evolving, plasma technology will claim its role in the years to follow as an essential surface modification technique.

Aknowledgments

This research has received funding from the European Research Council under the European Union's Seventh Framework Program (FP/2007-2013) / ERC Grant Agreement n. 279022.

References

[1] Jacobs, T., et al., Plasma modification of PET foils with different crystallinity. *Surface and Coatings Technology*, 2011. 205, Supplement 2(0): p. S511-S515.

[2] Tendero, C., et al., Atmospheric pressure plasmas: A review. *Spectrochimica Acta Part B: Atomic Spectroscopy*, 2006. 61(1): p. 2-30.

[3] De Geyter, N., Plasma modification of polymer surfaces in the subatmospheric pressure range, in applied physics2008, ghent. p. 271.

[4] De Geyter, N., et al., Increasing the hydrophobicity of a PP film using a helium/CF4 DBD treatment at atmospheric pressure. *Plasma Chemistry and Plasma Processing,* 2008. 28(2): p. 289-298.

[5] Ren, C.-S., et al., Surface modification of PE film by DBD plasma in air. *Applied surface science,* 2008. 255(5): p. 3421-3425.

[6] Berthiaume, F., T.J. Maguire, and M.L. Yarmush, Tissue engineering and regenerative medicine: history, progress, and challenges. *Annual review of chemical and biomolecular engineering*, 2011. 2: p. 403-430.

[7] Conrads, H. and M. Schmidt, Plasma generation and plasma sources. *Plasma Sources Science and Technology,* 2000. 9(4): p. 441.

[8] Schutze, A., et al., The atmospheric-pressure plasma jet: a review and comparison to other plasma sources. *Plasma Science, IEEE Transactions* on, 1998. 26(6): p. 1685-1694.

[9] Hopwood, J., Review of inductively coupled plasmas for plasma processing. *Plasma Sources Science and Technology*, 1992. 1(2): p. 109.

[10] Kogelschatz, U., B. Eliasson, and W. Egli, From ozone generators to flat television screens: history and future potential of dielectric-barrier discharges. *Pure and Applied Chemistry,* 1999. 71(10): p. 1819-1828.

[11] Laroussi, M. and T. Akan, Arc-Free Atmospheric Pressure Cold Plasma Jets: A Review. *Plasma Processes and Polymers*, 2007. 4(9): p. 777-788.

[12] Larsson Wexell, C., et al., Bone Response to Surface-Modified Titanium Implants: Studies on the Early Tissue Response to Implants with Different Surface Characteristics. *International Journal of Biomaterials*, 2013. 2013.

[13] Morra, M. and C. Cassinelli, Evaluation of surface contamination of titanium dental implants by LV-SEM: comparison with XPS measurements. *Surface and Interface Analysis,* 1997. 25(13): p. 983-988.

[14] Cools, P., et al., Surface Analysis of Titanium Cleaning and Activation Processes: Non-thermal Plasma Versus Other Techniques. *Plasma Chemistry and Plasma Processing*: p. 1-16.

[15] Constantine, C., et al., Plasma etching of III–V semiconductors in CH 4/H 2/Ar electron cyclotron resonance discharges. *Journal of Vacuum Science & Technology B: Microelectronics and Nanometer Structures*, 1990. 8(4): p. 596-606.

[16] Fracassi, F., et al., Plasma assisted dry etching of cobalt silicide for microelectronics applications. *Journal of The Electrochemical Society*, 1996. 143(2): p. 701-707.

[17] Lee, J., J. Hong, and S. Pearton, Etching of InP at$\gtrsim$ 1 μm/min in Cl 2/Ar plasma chemistries. *Applied physics letters*, 1996. 68(6): p. 847-849.

[18] Chang, J.P., et al., Kinetic study of low energy argon ion-enhanced plasma etching of polysilicon with atomic/molecular chlorine. *Journal of Vacuum Science & Technology A: Vacuum, Surfaces, and Films*, 1997. 15(4): p. 1853-1863.

[19] Washburn, N.R., et al., High-throughput investigation of osteoblast response to polymer crystallinity: influence of nanometer-scale roughness on proliferation. *Biomaterials,* 2004. 25(7): p. 1215-1224.

[20] Khorasani, M.T., H. Mirzadeh, and S. Irani, Plasma surface modification of poly (l-lactic acid) and poly (lactic-co-glycolic acid) films for improvement of nerve cells adhesion. *Radiation Physics and Chemistry*, 2008. 77(3): p. 280-287.

[21] Mattioli, S., J. Kenny, and I. Armentano, Plasma surface modification of porous PLLA films: analysis of surface properties and in vitro hydrolytic degradation. *Journal of Applied Polymer Science*, 2012. 125(S2): p. E239-E247.

[22] Jacobs, T., et al., Plasma surface modification of polylactic acid to promote interaction with fibroblasts. Journal of Materials Science: *Materials in Medicine*, 2013. 24(2): p. 469-478.

[23] Owens, D.K. and R. Wendt, Estimation of the surface free energy of polymers. *Journal of applied polymer science*, 1969. 13(8): p. 1741-1747.

[24] Watts, J.F. and J. Wolstenholme, An introduction to surface analysis by XPS and AES. An Introduction to Surface Analysis by XPS and AES, by John F. Watts, John Wolstenholme, pp. 224. ISBN 0-470-84713-1. Wiley-VCH, May 2003., 2003. 1.

[25] Siow, K.S., et al., Plasma Methods for the Generation of Chemically Reactive Surfaces for Biomolecule Immobilization and Cell Colonization-A Review. *Plasma processes and polymers*, 2006. 3(6-7): p. 392-418.

[26] Batich, C.D., Chemical derivatization and surface analysis. *Applied Surface Science,* 1988. 32(1): p. 57-73.

[27] Holländer, A., et al., Distinguishing Surface OH and NHx Using TFAA Derivatization and XPS. *Plasma Processes and Polymers*, 2008. 5(4): p. 345-349.

[28] Yang, J., J. Bei, and S. Wang, Enhanced cell affinity of poly (D, L-lactide) by combining plasma treatment with collagen anchorage. *Biomaterials*, 2002. 23(12): p. 2607-2614.

[29] Zhao, Y., et al., Effects of gas atmospheres on poly (lactic acid) film in acrylic acid plasma treatment. *Applied Surface Science*, 2013. 283: p. 181-187.

[30] Ferreira, B., et al., Plasma surface treatments of poly (l-lactic acid)(PLLA) and poly (hydroxybutyrate-co-hydroxyvalerate)(PHBV). *Materials Science and Engineering*: C, 2009. 29(3): p. 806-813.

[31] Yang, F., et al., Manufacturing and morphology structure of polylactide-type microtubules orientation-structured scaffolds. *Biomaterials*, 2006. 27(28): p. 4923-4933.

[32] Yang, J., et al., Fabrication and surface modification of macroporous poly (L-lactic acid) and poly (L-lactic-co-glycolic acid)(70/30) cell scaffolds for human skin fibroblast cell culture. *Journal of biomedical materials research*, 2002. 62(3): p. 438-446.

[33] Xu, Z.X., et al., Amide-linkage formed between ammonia plasma treated poly (D, L-lactide acid) scaffolds and bio-peptides: Enhancement of cell adhesion and osteogenic differentiation in vitro. *Biopolymers,* 2011. 95(10): p. 682-694.

[34] Slepičková Kasálková, N., et al., Biocompatibility of plasma nanostructured biopolymers. *Nuclear Instruments and Methods in Physics Research Section B: Beam Interactions with Materials and Atoms*, 2013. 307: p. 642-646.

[35] Slepičková Kasálková, N., et al., Surface changes of biopolymers PHB and PLLA induced by Ar< sup>+</sup> plasma treatment and wet etching. *Nuclear Instruments and Methods in Physics Research Section B: Beam Interactions with Materials and Atoms,* 2014.

[36] Wan, Y., et al., Cell adhesion on gaseous plasma modified poly-(L-lactide) surface under shear stress field. *Biomaterials*, 2003. 24(21): p. 3757-3764.

[37] Gomathi, N., A. Sureshkumar, and S. Neogi, RF plasma-treated polymers for biomedical applications. *CURRENT SCIENCE-BANGALORE-*, 2008. 94(11): p. 1478.

[38] Koh, H., et al., Enhancement of neurite outgrowth using nano-structured scaffolds coupled with laminin. *Biomaterials,* 2008. 29(26): p. 3574-3582.

[39] Shakhssalim, N., et al., Bladder tissue engineering using biocompatible nanofibrous electrospun constructs: feasibility and safety investigation. *Urology journal*, 2012. 9(1): p. 410-419.

[40] Kabiri, M., et al., Neural differentiation of mouse embryonic stem cells on conductive nanofiber scaffolds. *Biotechnology letters*, 2012. 34(7): p. 1357-1365.

[41] Park, K., et al., Acrylic acid-grafted hydrophilic electrospun nanofibrous poly (L-lactic acid) scaffold. *Macromolecular Research*, 2006. 14(5): p. 552-558.

[42] Armentano, I., et al., Role of PLLA plasma surface modification in the interaction with human marrow stromal cells. *Journal of applied polymer science*, 2009. 114(6): p. 3602-3611.

[43] Hirotsu, T., et al., Plasma surface treatments of melt-extruded sheets of poly (L-lactic acid). *Polymer Engineering & Science*, 2002. 42(2): p. 299-306.

[44] Yamaguchi, M., et al., Surface modification of poly (L-lactic acid) affects initial cell attachment, cell morphology, and cell growth. *Journal of Artificial Organs*, 2004. 7(4): p. 187-193.

[45] Riccardi, C., et al., Surface modification of poly (ethylene terephthalate) fibers induced by radio frequency air plasma treatment. *Applied surface science*, 2003. 211(1): p. 386-397.

[46] Alves, C., et al., Plasma surface modification of poly (D, L-lactic acid) as a tool to enhance protein adsorption and the attachment of different cell types. *Journal of Biomedical Materials Research Part B: Applied Biomaterials*, 2008. 87(1): p. 59-66.

[47] Jiao, Y., J. Xu, and C. Zhou, Effect of Ammonia Plasma Treatment on the Properties and Cytocompatibility of a Poly (L-Lactic Acid) Film Surface. *Journal of Biomaterials Science, Polymer Edition,* 2012. 23(6): p. 763-777.

[48] Sarapirom, S., et al., Effect of surface modification of poly (lactic acid) by low-pressure ammonia plasma on adsorption of human serum albumin. *Applied Surface Science,* 2014.

[49] Yang, J., J. Bei, and S. Wang, Improving cell affinity of poly (D, L-lactide) film modified by anhydrous ammonia plasma treatment. *Polymers for Advanced Technologies,* 2002. 13(3-4): p. 220-226.

[50] Cheng, Q., et al., Plasma surface chemical treatment of electrospun poly (L-lactide) microfibrous scaffolds for enhanced cell adhesion, growth, and infiltration. *Tissue Engineering Part A,* 2013. 19(9-10): p. 1188-1198.

[51] Chu, C.F., et al., Enhanced growth of animal and human endothelial cells on biodegradable polymers. *Biochimica et Biophysica Acta (BBA)-General Subjects,* 1999. 1472(3): p. 479-485.

[52] Sipehia, R., X-ray photoelectron spectroscopy studies, surface tension measurements, immobilization of human serum albumin, human fibrinogen and human fibronectin onto ammonia plasma treated surfaces of biomaterials useful for cardiovascular implants and artificial cornea implants. *Artificial Cells, Blood Substitutes and Biotechnology*, 1993. 21(5): p. 647-658.

[53] Inagaki, N., et al., Surface modification and degradation of poly (lactic acid) films by Ar-plasma. *Journal of adhesion science and technology*, 2002. 16(8): p. 1041-1054.

[54] Slepička, P., et al., Surface modification of biopolymers by argon plasma and thermal treatment. *Plasma Processes and Polymers*, 2012. 9(2): p. 197-206.

[55] Ding, Z., et al., Immobilization of chitosan onto poly-L-lactic acid film surface by plasma graft polymerization to control the morphology of fibroblast and liver cells. *Biomaterials,* 2004. 25(6): p. 1059-1067.

[56] Slepička, P., et al., Cytocompatibility of plasma and thermally treated biopolymers. *Journal of Nanomaterials*, 2013. 2013: p. 10.

[57] Song, W., et al., Bioinspired degradable substrates with extreme wettability properties. *Advanced Materials*, 2009. 21(18): p. 1830-1834.

[58] Inagaki, N., plasma surface modification and plasma polymerization1996: technomic publishing company inc.

[59] Ho, M.H., et al., Promotion of cell affinity of porous PLLA scaffolds by immobilization of RGD peptides via plasma treatment. *Macromolecular bioscience*, 2006. 6(1): p. 90-98.

[60] Chaiwong, C., et al., Effect of plasma treatment on hydrophobicity and barrier property of polylactic acid. *Surface and Coatings Technology*, 2010. 204(18–19): p. 2933-2939.

[61] Poncin-Epaillard, F., O. Shavdina, and D. Debarnot, Elaboration and surface modification of structured poly (l-lactic acid) thin film on various substrates. *Materials Science and Engineering*: C, 2013. 33(5): p. 2526-2533.

[62] Boonyawan, D., et al., Characterization and antimicrobial properties of fluorine-rich carbon films deposited on poly(lactic acid). *Surface and Coatings Technology*, 2011. 205, Supplement 2(0): p. S552-S557.

[63] Roy, R.K., et al., Hemocompatibility of surface-modified, silicon-incorporated, diamond-like carbon films. *Acta Biomaterialia*, 2009. 5(1): p. 249-256.

[64] Demina, T., et al., DC discharge plasma modification of chitosan/gelatin/PLLA films: Surface properties, chemical structure and cell affinity. *Surface and Coatings Technology*, 2012. 207: p. 508-516.

[65] Khorasani, M., H. Mirzadeh, and S. Irani, Comparison of fibroblast and nerve cells response on plasma treated poly (L-lactide) surface. *Journal of applied polymer science,* 2009. 112(6): p. 3429-3435.

[66] De Geyter, N., et al., Treatment of polymer films with a dielectric barrier discharge in air, helium and argon at medium pressure. *Surface and Coatings Technology*, 2007. 201(16): p. 7066-7075.

[67] Morent, R., et al., Influence of discharge atmosphere on the ageing behaviour of plasma-treated polylactic acid. *Plasma Chemistry and Plasma Processing*, 2010. 30(4): p. 525-536.

[68] De Geyter, N., et al., Plasma modification of polylactic acid in a medium pressure DBD. *Surface and Coatings Technology*, 2010. 204(20): p. 3272-3279.

[69] Hergelová, B., et al., *Plasma Assisted Activation of Polylactic Acid Surface*.

[70] Vergne, C., et al., Modifications of the polylactic acid surface properties by DBD plasma treatment at atmospheric pressure. *Journal of Engineering Materials and Technology*, 2011. 133(3): p. 030903.

[71] Barradas, A., et al., Surface modifications by gas plasma control osteogenic differentiation of MC3T3-E1 cells. *Acta biomaterialia*, 2012. 8(8): p. 2969-2977.

[72] Nakagawa, M., et al., Improvement of cell adhesion on poly (L-lactide) by atmospheric plasma treatment. *Journal of Biomedical Materials Research Part A,* 2006. 77(1): p. 112-118.

[73] Teraoka, F., M. Nakagawa, and M. Hara, Surface modification of poly (L-lactide) by atmospheric pressure plasma treatment and cell response. *Dental materials journal,* 2006. 25(3): p. 560-565.

[74] Yang, Y.W., et al., Fast incorporation of primary amine group into polylactide surface for improving C2C12 cell proliferation using nitrogen-based atmospheric-pressure

plasma jets. *Journal of Biomedical Materials Research Part A*, 2014. 102(1): p. 160-169.

[75] Safinia, L., et al., Atmospheric plasma treatment of porous polymer constructs for tissue engineering applications. *Macromolecular bioscience*, 2007. 7(3): p. 315-327.

[76] Ni, H.-C., et al., The use of air plasma in surface modification of peripheral nerve conduits. *Acta biomaterialia*, 2010. 6(6): p. 2066-2076.

[77] Jordá-Vilaplana, A., et al., Surface modification of polylactic acid (PLA) by air atmospheric plasma treatment. *European Polymer Journal*, 2014.

[78] Chen, J., S. Lee, and I. Chiang, Plasma-treated and collagen-coated plla nanofiber membrane for cartilage tissue engineering. *NSTI*. 2(5): p. 100-103.

[79] Yokoyama, Y., A. Oyane, and A. Ito, Biomimetic coating of an apatite layer on poly (L-lactic acid); improvement of adhesive strength of the coating. *Journal of Materials Science: Materials in Medicine,* 2007. 18(9): p. 1727-1734.

[80] Park, K., et al., Effect of surface-activated PLLA scaffold on apatite formation in simulated body fluid. *Journal of Bioactive and Compatible Polymers*, 2010. 25(1): p. 27-39.

[81] Barry, J.J., et al., Using Plasma Deposits to Promote Cell Population of the Porous Interior of Three-Dimensional Poly (D, L-Lactic Acid) Tissue-Engineering Scaffolds. *Advanced Functional Materials*, 2005. 15(7): p. 1134-1140.

[82] Brunette, D.M. and e. al., Titanium in medicine: material science, surface science, engineering, biological responses and medical applications. Berlin, Springer-Verlag, 2001.

[83] García, C.G., et al., Poly (L-lactide) Substrates with Tailored Surface Chemistry by Plasma Copolymerisation of Acrylic Monomers. *Plasma Processes and Polymers*, 2009. 6(3): p. 190-198.

[84] Da Ponte, G., et al., Trends in surface engineering of biomaterials: atmospheric pressure plasma deposition of coatings for biomedical applications. *European Physical Journal-Applied Physics,* 2011. 56(2): p. 7.

[85] Da Ponte, G., et al., Atmospheric Pressure Plasma Deposition of Poly Lactic Acid-Like Coatings with Embedded Elastin. *Plasma Processes and Polymers*, 2014. 11(4): p. 345-352.

[86] available as free-source material at: http://upload.wikimedia.org/wikipedia/commons /archive/1/18/20110306112911!Electric-discharge-regimes-illustrated.jpg

[87] Reprinted from Surface and Coatings Technology, 2007. 201(16): p. 7066-7075 De Geyter, N., Morent R., Leys C., Gengembre L., Payen E., Treatment of polymer films with a dielectric barrier discharge in air, helium and argon at medium pressure., Copyright © 2007, with permission from Elsevier.

[88] Reprinted from Journal of Applied Polymer Science, 2012. 125(S2): p. E239-E247, S. Mattioli,J. M. Kenny,I. Armentano, Plasma surface modification of porous PLLA films: Analysis of surface properties and in vitro hydrolytic degradation, Copyright © 2012 Wiley Periodicals, Inc.

[89] Reprinted from Biopolymers, 2011. 95(10): p. 682-694, Zi-Xing Xu,Tao Li,Zhao-Ming Zhong,Ding-Sheng Zha,Song-Hui Wu,Fu-Qiang Liu,Wen-De Xiao,Xiao-Rui Jiang,Xin-Xin Zhang,Jian-Ting Chen, Amide-linkage formed between ammonia plasma treated poly (D, L-lactide acid) scaffolds and bio-peptides: Enhancement of cell adhesion and osteogenic differentiation in vitro, Copyright © 2011 Wiley Periodicals, Inc.

[90] Reprinted from Tissue Engineering Part A, 2013. 19(9-10): p. 1188-1198, Qian Cheng, Benjamin Li-Ping Lee, Kyriakos Komvopoulos, et al., Plasma surface chemical treatment of electrospun poly (L-lactide) microfibrous scaffolds for enhanced cell adhesion, growth, and infiltration, Copyright © 2013, Mary Ann Liebert, Inc.

[91] Reprinted from Advanced Materials, 2009. 21(18): p. 1830-1834, Wenlong Song,Diana D. Veiga,Catarina A. Custódio,João F. Mano, Bioinspired degradable substrates with extreme wettability propertie, Copyright © 2009 WILEY-VCH Verlag GmbH & Co. KGaA, Weinheim.

[92] Reprinted from Journal of Applied Polymer Science, 2003. 89: P. 96-103, N. Inagaki,K. Narushima,S. K. Lim, Effects of aromatic groups in polymer chains on plasma surface modification, Copyright © 2003 Wiley Periodicals, Inc.

[93] Reprinted from Advanced Functional Materials, 2005. 15(7): p. 1134-1140, J. J. A. Barry,M. M. C. G. Silva,K. M. Shakesheff,S. M. Howdle,M. R. Alexander, Using Plasma Deposits to Promote Cell Population of the Porous Interior of Three-Dimensional Poly (D, L-Lactic Acid) Tissue-Engineering Scaffolds, Copyright © 2005 WILEY-VCH Verlag GmbH & Co. KGaA, Weinheim.

In: New Developments in Polylactic Acid Research
Editor: Courtney Winthrop
ISBN: 978-1-63463-054-2

Chapter 4

Polylactic Acid Composites

W. Jia and R. H. Gong[*]
Textiles, School of Materials, the University of Manchester, Manchester, UK

Abstract

Polylactic acid (PLA) is well known as a fully biodegradable and sustainable polymer. Due to its high price in the past, its initial applications are mainly biomedical materials, such as bone fixation devices and implant meshes. Due to the increasing awareness of environment protection and sustainable development, the "green" polymer, PLA, has been developing rapidly. The applications of PLA have expanded into plastics, textiles, composites, etc. Since most of the traditional composite materials, such as carbon fibre or glass fibre reinforced epoxy composites, are composed of two distinct fossil fuel based components, they are not eco-friendly and difficult to recycle. Development of biodegradable composites is becoming more attractive. This chapter reviews the development of PLA based biodegradable composites, includes natural fibre reinforced PLA and PLA fibre reinforced composite systems. The main manufacturing processes will be described along with key material and process parameters. The future of these PLA based composites will be explored.

Introduction

Polylactic acid (PLA) is a biodegradable aliphatic polyester derived from 100% renewable resources. The price of PLA was relatively high in the past compared to commodity thermoplastic polymers such as polyethylene and polypropylene; thus, its initial applications were mainly biomedical materials. However, the price of PLA can be decreased by using crop resides from corn plants, such as the stems, straw, husks and leaves of the corn [1]. Due to the rapid development of eco-friendly materials, PLA can now be manufactured in large-scale, and be routinely found in a broad range of applications including apparel, food packaging, diapers, wipes, and composites.

[*] Email: Hugh.Gong@manchester.ac.uk.

PLA has excellent biocompatibility, biodegradability and thermoplastic processability. PLA can be easily processed by various conventional processing techniques, such as extrusion, injection moulding, compression moulding, and blow moulding. However, the poor mechanical properties and low thermal stability of PLA polymer restrict its applications as high performance materials. In order to spread the uses of PLA, foreign fillers (fibres) are introduced into the PLA to create a composite with improved material properties. Hence, PLA based composites have drawn intensive attentions around the world. This chapter reviews the recent developments in the manufacturing and mechanical property evaluation of PLA based composites.

Polylactic Acid

Poly (lactic acid) is not a new material. It was developed in 1932 by Wallace Carothers, a scientist working for Du Pont Company [2]. In recent years it has been widely developed and manufactured on a large-scale by NatureWorks LLC which markets the fibre form of the PLA polymer under the "INGEO™" brand.

PLA is a linear aliphatic thermoplastic polyester, commonly made from fully renewable resources, such as wheat, corn and sugar cane and other products which contain rich carbohydrates [3]. There are mainly two ways to produce PLA polymer. They are direct condensation polymerization of lactic acid and ring-opening polymerization that occurs via lactide as a cyclic intermediate dimer [1, 4]. The ring-opening polymerization usually gives longer molecules (higher molecular weight) than the direct polycondensation because it is difficult to remove the water and impurities in polycondensation [4-6].

Due to the chirality of lactic acid, it exists as two stereosiomers, which are L- and D-lactic acids. The chemical structures of Poly (L-lactic) acid and poly (D-lactic) acid are shown in Figure 1. The characteristics of PLA, such as molecular weight, crystallinity, thermal properties, can be changed by varying the amount and sequence of L- and D-lactic acid units [4, 7]. The crystalline PLA polymer usually contains high proportion of L-lactic acid unit, while the amorphous PLA polymer is produced by a higher proportion of D-lactic acid unit (more than 15%) in the polymer chain [4]. The melting point of PLA polymer is variable in the range from 130°C to 230°C by changing the ratio and distribution of L- and D-isomers in the polymer backbone. Ikada et al., [8] found that the highest melting point of 230°C of PLA polymer can be achieved by blending PLLA and PDLA to create a stereocomplex polymer.

Similar to most of thermoplastic polymers, the mechanical properties of PLA is highly depended on the crystallinity, molecular weight and orientation. Semi-crystalline PLA is characterized by good mechanical properties, with tensile strength of 50-70 MPa, tensile elastic modulus of 3 - 4 GPa, elongation at break of 2-10%, flexural strength of 50-120 MPa, and flexural modulus of 3.5-5 GPa [9, 10].

The mechanical properties of PLA can be improved efficiently by drawing the PLA into highly oriented fibre. The mechanical properties of the PLA fibre are determined by a variety of spinning parameters such as draw ratio, roll speed etc. The tensile strength, elastic modulus of PLA fibre can be developed up to 450 MPa (about 36 cN/tex, based on bulk density of 1.25 g/cm^3), 5.7 GPa (draw ratio of 4), respectively [11, 12]. PLA fibre is also very extendable. The elongation at break can reach 160% by using a spinning speed of 1000 m/min) [11, 12].

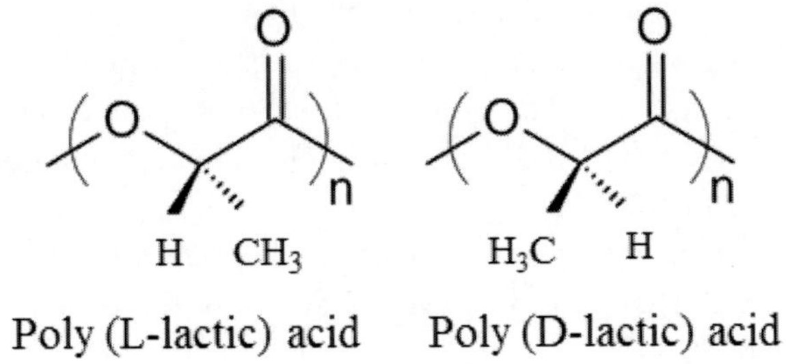

Figure 1. Chemical structures of PLLA and PDLA.

PLA has excellent biodegradability and compostability despite being a synthetic polymer. PLA can be degraded by hydrolysis of the ester bond without any enzyme to catalyze this hydrolysis or by various microorganisms under controlled conditions [13, 14]. The degradation rate is influenced by ordered molecular structure (crystallinity), molecular weight, surface area, hydrophilic and hydrophobic properties [13]. Because the crystalline structure is more resistant to degradation then the amorphous one, the higher the crystallinity is, the lower the degradation rate will be. In addition, the degradation of PLA depends on the degrading environmental condition such as temperature and humidity. Under the natural environment, PLA has a degradation time from six months to two years which is much shorter than that of polyethylene which is 500-1000 years [14]. However, the degradation time of PLA can be shortened to 3-4 weeks when PLA is in composts with microorganisms in appropriate environment [15].

Composites and Biocomposites

Composite is defined as natural or engineered materials usually made up of two or more constituents with different physical or chemical properties that maintain distinct appearances on a macroscopic or microscopic scale in the final product [16]. Unlike a single material, which have all the excellent properties on its own, a composite is designed to have the desired balance of properties that are contributed by each component, based on the particular applications [17]. In a modern reinforced composite, a stiff and strong component, known as the reinforcement (fibre), is usually embedded in a softer continuous component called the matrix [16]. The function of the reinforcement in a composite is to carry the applied load and provide structural properties such as stiffness, strength, conductivity and thermal stability. The matrix keeps the reinforcements in place and at proper orientation, transfers the load to the reinforcements, protects them from mechanical damages, provides rigidity and shape to the structure and manages the transverse properties of the composites. Depending on the nature of the matrixes, composite can be mainly classified into three types: polymer, metal and ceramic matrix composites. Reinforcements can be in the form of long or short fibres, particulates, and monofilaments.

Composites have been used in a wide range of industrial applications for several decades. Large-scale applications include various land and sea transportation, such as aircrafts and vehicles; small-scale applications cover sport equipment and biomedical materials, such as tennis rackets and bond fixation devices. However, most commercial composites are fossil fuel based synthetic materials which have adverse impacts on the environment due to the limited amount of fossil fuel and the problems of disposal at the end of the product life cycle. Furthermore, with the growth of the environmental awareness and consequent upon a series of new regulations, the industries are being forced to search for eco-friendly materials to replace the existing fossil fuel based polymeric materials, in order to help preserve and protect the environment. Therefore, interest has grown in the development of biodegradable composites (biocomposites).

Biocomposites have a broad range and can be partial or completely biodegradable [18, 19]. The classification of biocomposites is shown in Figure 2. The partial biodegradable composites are based on biofibres as reinforcement and traditional non-biodegradable thermoplastic (polypropylene/polyethylene) or thermoset (epoxy/unsaturated polyester) polymers as matrix. If the matrixes are biodegradable polymers, which can be derived from renewable or petro-based resources, the biocomposites are completely biodegradable. The renewable polymers include starch plastics, soy plastics, cellulosic acetate and PLA; the petro-based synthetic polymers include aliphatic polyester, aliphatic-aromatic polyester, polyesteramides, and polyvinyl alcohols [18-20].

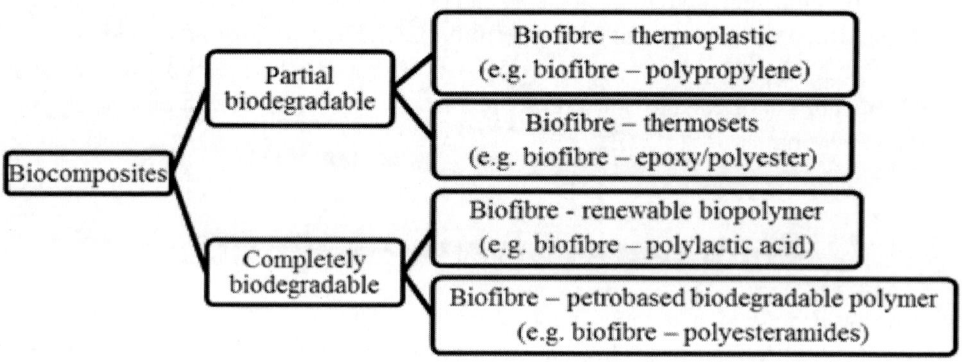

Figure 2. Classification of biocomposites [18, 19].

In biocomposites, biofibres are mainly natural fibres, which can be grouped into plant fibres, animal fibres, and mineral fibres. The plant fibres cover wood fibres, bast (jute, flax, hemp, kenaf), leaf (sisal, abaca, and pineapple), straw/grass (bamboo, rice, wheat, and corn) fibres and seed/fruit (cotton and coconut coir) fibres [18-20]. Due to their low cost, low density, and comparable mechanical properties to glass fibre, natural fibres have been considered as feasible alternatives to the expensive glass, aramid carbon fibres in a composite, especially in automobile and construction industries [18-21]. An all-composite automobile is more than 50% lighter than a similar sized steel automobile and it could be even lighter by using natural fibre reinforced composites since natural fibres normally have lower density than synthetic fibres [21]. The comparison of mechanical properties between natural fibres and high-performance synthetic fibres are shown in Table 1. In general, mechanical properties of natural fibres are lower than those of synthetic fibres. However, the stiffness of hemp and ramie are similar or even higher than that of the glass fibre; in addition, the specific modulus (modulus/specific gravity) of natural fibres is comparable to that of glass fibre [20-22].

Table 1. Properties of natural fibres and high-performance fibres [20-22]

Fibre	Density (g/cm^3)	Tensile strength (MPa)	Young's modulus (GPa)	Elongation at break (%)
Jute	1.3-1.49	393-800	13-26.5	1.16-1.5
Flax	1.5	345-1500	27-39	2.7-3.2
Hemp	1.47	550-900	38-70	1.6-4
Kenaf	1.5-1.6	350-930	40-53	1.6
Ramie	1.5-1.6	400-938	6.4-128	1.2-3.8
Sisal	1.45	468-700	9.4-22	3-7
Coir	1.15-1.46	131-220	4-6	15-40
E-glass	2.5	2000-3500	70-73	2.5
S-glass	2.5	4570	86	2.8
Aramid	1.4	3000-3150	60-67	2.5-3.7
Carbon	1.78	4000-4800	230-425	1.4-1.8

Furthermore, it has been reported in several studies that 10-11 million vehicles in the United States need to be disposed annually after their end-use, and approximately 25% weight of the vehicles are discarded as waste, including plastics, composites, rubber and forms [18, 20]. The other advantage of using biocomposites to replace glass fibre composite is that biocomposites have good biodegradability; they can be simply consumed by landfill. In the last few decades, natural fibre reinforced composites have been used widely as interior and exterior material with multiple functions. They have been used as door panels, door trim panels, floor mats and automobile spare tire covers to reduce the weight for higher fuel efficiency and to enhance sustainability [18, 21].

Synthetic Fibres Reinforced PLA Composites

Synthetic fibres are commonly employed as reinforcement in a composite to provide structural properties due to their stronger mechanical properties and higher uniformity compared to natural fibres. Glass fibre is one of the favourite fibres in the composite industry because it has high strength to weight ratio, dimensional stability, resistance to heat moisture and corrosion, and good electrical insulation [23]. Some studies have showed the feasibility and advantages of the glass fibre reinforced PLA composites in bone fixture device and novel scaffolds for applications in tissue engineering [24-26]. A unidirectional cellular glass/PLA composite that was produced by filament winding and supercritical gas foaming demonstrated that the compression modulus increased significantly with the fibre volume fraction [24]. Short phosphate–based glass fibre nonwoven mat as reinforcement was used to fabricate a PLA composite via film-stacking followed by the hot compression method [25, 26]. The flexural modulus and strength of the short glass fibre reinforced PLA composite with 40% fibre volume fraction can reach up to 10 GPa and 120 MPa respectively, which match the requirements of flexural properties for cortical bond [26]. Figure 3 shows the good PLA impregnation of the unidirectional and non-woven structure glass fibres. In the comparison of the chopped glass fibre reinforced PLA and polypropylene (PP) composites with the same fibre volume fraction, glass/PLA composite shows higher tensile strength and modulus [27].

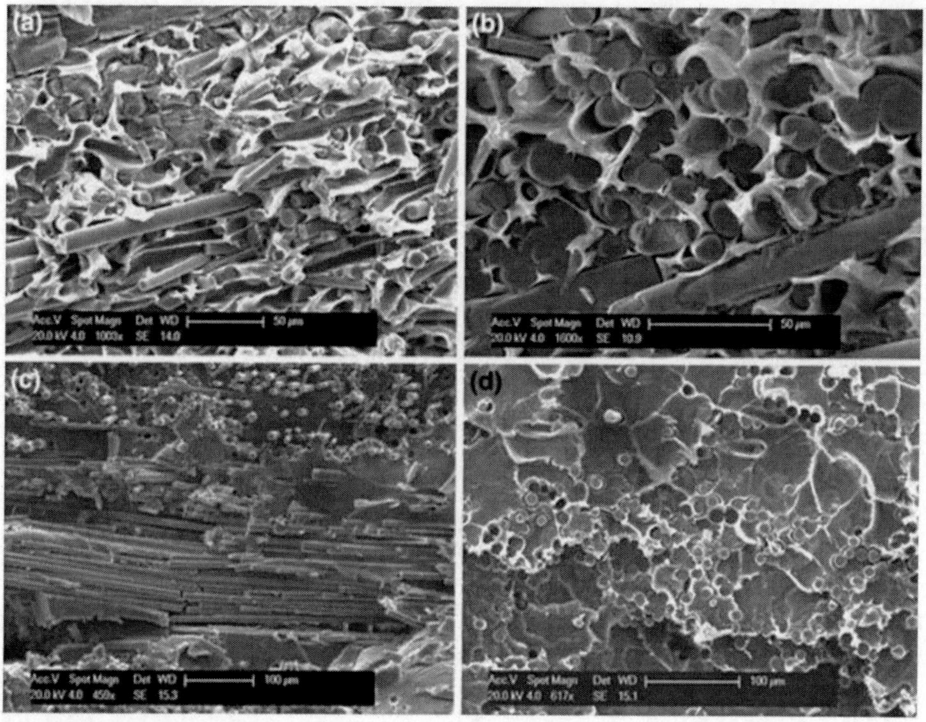

Figure 3. SEM images of non-woven structure with 30% (a) and 40% (b) fibre volume fraction and unidirectional (c) and (d) glass fibre [26].

Graupner et al. found Lyocell fibre, a man-made cellulose fibre, reinforced PLA composite has superior tensile strength, elongation at break and impact strength, compared to cotton, kenaf and hemp reinforced PLA [28]. Cordenka, another man-made cellulose fibre, is a rayon fibre usually used for reinforced ties [29]. Ganster and Fink reported that Cordenka - PLA composite (25% fibre weight fraction) increases significantly the tensile strength and modulus of neat PLA [30]. Bax and Müssig also found the impact strength, tensile strength and Young's modulus of Cordenka- PLA composites increase with the increase of fibre-mass proportion within 10-30%; and the composite impact strength is significantly higher than flax fibre reinforced PLA with the same fibre-mass proportion.

Natural Fibre Reinforced PLA Biocomposites

Bast Fibres Reinforced PLA Composites

Compared to synthetic fibres, bast fibres have lower density and higher biodegradability. Since they also have relatively high Young's modulus and tensile strength, research on bast fibre reinforced PLA composites have attracted significant attention. The bast fibres that are usually used for biodegradable composites are hemp [31-33], jute [34, 35], flax [29, 36-38] and kenaf [39-43].

Various manufacturing techniques are used to produce these composites. The most common manufacturing processes are injection moulding and compression moulding (using film-stacking). The injection moulding method can provide good degree of fibre wetting, because the fibre and PLA matrix can be mixed properly in an extruder [29, 33, 42-44]. The disadvantage of this method is that the reinforcement structure of the composite is limited to chopped short fibre. On the other hand, compression moulding can use a variety of reinforced structures, including short fibre mat [31, 34, 36, 38], chopped short fibre [32, 40], unidirectional filaments and textile fabric [39, 41]. The degree of fibre wetting is highly depended on the processing temperature, pressure and time. To improve matrix fusion and wettability of fibres, Khonder and the co-workers [35] used a tubular braiding technique to a micro-braid yarn. The micro-brad yarn was constituted by a jute yarn as the straight inserted axial reinforcement fibre and the PLA matrix fibre was braided around the reinforcing jute yarn. Then, the micro-braiding yarns were compression moulded to create a unidirectional composite.

The mechanical properties of bast fibre reinforced composites are found to be comparable to glass fibre reinforced PLA composites. They have a tensile strength of 80 MPa and an elastic modulus of 6.7 GPa [27]. Oksman et al. have found that the flax fibre (30-40 wt %) reinforced PLA composite has 44-53 MPa tensile strength and 7.3-8.3 GPa E-modulus, which are higher than or equivalent to those of PP composites [44]. The mechanical properties of flax-PLA composites with higher fibre content (40-60 wt %) have been reported in Alimuzzaman et al.'s study [38]. By changing the moulding temperature and time during heat compression, the highest tensile properties (tensile strength = 80.3 MPa, tensile modulus = 9.9 GPa) flexural (flexural strength = 138.5 MPa, flexural modulus = 9.9 GPa) are achieved by the 50 wt% flax composite at 180°C moulding temperature and 5 min moulding time.

The unidirectional kenaf-PLA composite with 70 % fibre content was developed in Ochi's study. The tensile and flexural strengths can reach 223 MPa and 254 MPa respectively [39]. The tensile modulus and strength of a short kenaf fibre –PLA with 30% fibre content can reach about 5.5 GPa and 52 MPa, and the flexure modulus and strength are 6 GPa and 85 MPa, respectively [43]. Kenaf-PLA composites also have been found to have high heat resistance and strength; thus, they can be used in the housing of electronic products [42].

The mechanical properties of PLA can be significantly improved by inserting jute fibre. Khonder et al. [35] developed unidirectional jute-PLA composites with different fibre volume fractions by using different moulding pressures and temperatures. The tensile strength and modulus increase with increasing fibre volume fraction; while the bending strength and modulus tend to decrease. The maximum tensile strength and modulus are approximately 78 MPa and 8.5 GPa; the maximum bending strength and modulus are 85 MPa and 7 GPa, respectively. Plackett et al. [34] also showed that the mechanical properties of jute-PLA composites are highly determined by the processing temperature. Due to the high melt viscosity of the PLA matrix, the degree of fibre wetting is very poor at relatively low temperature. The tensile strength and modulus of jute-PLA composite are enhanced apparently to 100 MPa and 9.5 GPa by increasing the processing temperature to 210 - 220°C.

Hu and Lim found [32] the mechanical properties of hemp-PLA composites peak at the fibre volume fraction about 50%. As fibre content increase, the tensile strength, elastic modulus and flexural strength decrease. A possible explanation for this is that the poor fibre wetting at high fibre content, which is resulted by the insufficient resin fraction to wet all the fibre surfaces. Sawpan et al. [45] successfully evaluated the fracture toughness of hemp-PLA composite by single-edge-notched-bending test. The fracture toughness of the composite is lower than the neat PLA and decreases with the increase of fibre content. This could be caused by the increased stress concentration and PLA matrix crystallinity [45] .

The poor interfacial adhesion is a well-known problem in natural fibre reinforced synthetic composites. Many studies have worked on the improvement of interfacial adhesion by fibre surface treatment, the addition of plasticizer and coupling agent. Three types of fibre treatments were used in Huda el al' s study, including aqueous alkaline solution, silane coupling agent, and a combination of both [40]. All these three treatments improve the flexural properties of kenaf-PLA composite compared to the untreated composite. This is because the coupling agent can enhance the interfacial bonding by increasing the cross linking degree in the interface and the fibre surface area [40]. The positive effect of the fibre treatment also includes the removal of fibre surface wax and pectin [46]. The alkali and silane fibre treatments also show significant improvement on the mechanical properties of hemp-PLA composites in several studies [31, 33]. However, Oksman et al. reported that the triacetin plasticizer cannot improve the impact properties of the flax-PLA composite and the fibre mechanical properties are significantly decreased [44].

Leaf Fibres Reinforced PLA Composites

Sisal fibre is extracted from the leaves of the Agave sisalana plant with high cellulose content and low microfibril angle [46]. Since it has high tensile strength (700 MPa) and modulus (22 GPa), it has become one of the most commonly used leaf fibres in the composite industry. Several studies have worked on the manufacturing of sisal fibre reinforced PLA composites and the improvement in their mechanical properties. Short sisal fibre reinforced PLA composites with various fibre content (10 - 40 %) were produced by melt blending followed by compression in Wu's study [47]. The tensile stress at break of the sisal-PLA composite decreases with the increase of fibre content. It was also found that the sisal–modified PLA (acrylic acid-grafted PLA) composites have significantly higher tensile stress at break than the unmodified PLA at the full range of fibre contents. This could be due to the improvement in the interfacial adhesion through covalent linkages between fibre and matrix or hydophobization of the cellulose surface [46]. The mechanical properties of a long sisal fibre reinforced PLA composite were investigated in Prajer and Ansell's study [46]. The caustic soda fibre treatment was used to enhance the interfacial adhesion between sisal and PLA, in order to improve the tensile and flexural strength and modulus. The highest flexural strength (286 MPa) and modulus (22GPa) are achieved by the treated sisal fibre (60 vol%) - PLA composite; the 50 vol% treated sisal-PLA has tensile strength of 205 MPa and modulus of 12 GPa.

Transcrystallization is known as a columnar layer along the polymer-fibre interface and is resulted by heterogeneous nucleation. The improvement on the mechanical properties of the composites by the formation of the transcrystallization is controversial; some researchers found positive effect [48] and some showed negative or no effect [49]. Wang et al. [50] found both alkali and silane treated sisal fibres have a nucleating ability to form the transcrystallize in PLA matrix but no influence on the transcrystallization of the sisal fibre compared to the untreated fibre.

Bajpai et al. [51] evaluated the friction and wear characteristics of the sisal, Grewia optiva and nettle fibre reinforced PLA composites by a pin-on-disc rotating type friction and wear tribometer. The wear performance of the PLA matrix is enhanced by the addition of the natural fibres. The sisal-PLA composite shows the highest specific wear rate at 1 m/s sliding speed among the three composites [51].

Banana fibre–PLA composites were developed using melt blending followed by compression [52]. The tensile, flexural and impact properties of silane treated and untreated banana fibre-PLA were assessed. Both treated and untreated composites have higher tensile and flexural properties but lower impact strength than the neat PLA matrix. Moreover, the treated banana fibre-PLA shows higher tensile and flexural strength, and elongation at break compared to the untreated composite. In the study of Asaithambi et al., modified and unmodified banana/sisal fibre reinforced PLA hybrid composites were fabricated by a twin-screw extruder and subsequent injection moulding [53]. The fibre modification is done by an alkali treatment followed by benzoyl peroxide treatment to remove hemicellulose and allied impurities. The modified fibre–PLA composite shows significantly higher tensile and flexural strength and modulus as well as impact strength, compared to the virgin PLA and unmodified fibre-PLA composite.

The mechanical properties of three types of treated abaca fibre-PLA composites were investigated by Shibata and co-workers [54]. The flexural modulus of all the treated and untreated abaca fibre- PLA composite are increased as the fibre content increases from 5% to 20%. There is no obvious difference in flexural properties between the untreated and the three treated fibres-PLA composite, although the SEM images show the improvement in interfacial adhesion in the cases of butyric anhydride and acetic anhydride treated abaca–PLA composites.

Wood Fibres Reinforced PLA Composites

Wood fibre has been used as reinforcement in many polymers such as polypropylene and polybutylene succinate to improve the mechanical properties and reduce the cost, Maple hardwood fibre reinforced PLA composites (fibre content 10-50 wt%) were produced by melt blending and subsequent injection moulding in the study of Way et al. [55]. The addition of wood fibre shows significant improvement on the mechanical properties compared to the neat PLA matrix. In particular, the tensile modulus (2.5 GPa) and flexural modulus (8.22 GPa) of the 50 wt% wood fibre-PLA composite were 98% and 123% higher than those of the neat PLA.

Pine wood flour (20% and 40% fibre content) reinforced PLA composite has been reported in Pilla et al.'s study [56]. The wood flour-PLA composite was fabricated by melt blending followed by injection moulding. The addition of pine wood flour increases the tensile modulus and maintains the tensile strength, but decreases the toughness and strain at break of the PLA matrix. The silane coupling agent did not offer significant effect on the properties of the pine wood flour–PLA. Hrabalove et al. [57] produced wood flour-PLA composites with various fibre content up to 50% by melt blending followed by compression. They found that the fibre-matrix compatibility and properties of the wood flour-PLA composite can be optimised by suitable crystalline structure, processing parameters and fibre content.

Pulp fibres, due to their attractive mechanical properties, high property reproducibility and abundant material supply, have been used successfully in the composite industry [58]. In Du et al.'s study [58], hardwood and softwood pulp fibre sheets were employed as reinforcement in PLA composites. 30-50 wt% pulp fibres-PLA composites were fabricated by combining wet-laid paper-making process with film-stacking process. All the composites show significantly higher tensile modulus and strength than the PLA matrix. Besides, the calculated fibre length efficiency factor in the study for composite tensile strength is very close to continuous random fibre-reinforced composites.

Straw and Grass Fibres Reinforced PLA Composites

Bamboo is an abundant plant in South-eastern countries and is one of the fastest growing plants. As a result, bamboo fibre has high availability and relatively low cost. Moreover, bamboo fibre has high tensile modulus and elongation at break; its stiffness and specific strength are comparable to those of glass fibre. Thus, many studies have been carried out on bamboo fibre reinforced PLA composites [59-63]. Various short bamboo fibre–PLA composites (10-50 wt% fibre content) were produced by extrusion and subsequent injection or compression moulding and film-stacking [59-61]. Due to poor interfacial bonding between bamboo and PLA composites, these studies used a variety of fibre surface treatments and compatibilizer to improve the interfacial adhesion [59-61]. All of these treatments show significant improvement on the mechanical properties of bamboo fibre–PLA composites. The mechanical properties of long bamboo fibre-PLA composites were reported in Rawi et al.'s study [62, 63]. Plain woven bamboo fabric reinforced PLA composites were fabricated by film-stacking at various processing conditions. Optimised processing parameters of 160 °C temperature, 1.05 MPa pressure and 3 minute time were defined. The optimised processing parameters provide tensile modulus and strength of 5.9 GPa and 80.64 MPa, flexural modulus and strength of 4.5 GPa and 143 MPa in the stronger, warp direction [62]. Compared to bamboo fabric-PP composite, the PLA composite offer more significant improvement on impact strength [63].

A large amount of rice or wheat straw and husks is incinerated and discarded every year as agriculture waste. Because of the low content of cellulose in the rice straw fibre, it is difficult to achieve in good composite properties and there is not much research on it [64]. However, rice straw fibre does have potential as reinforcement in a biodegradable PLA composite because of its low cost and improvable physico-chemical properties by various treatments. Qin et al. [64] found the tensile strength of the rice straw fibre–PLA composite can be enhanced from 24 MPa to 30 MPa by using poly(butyl acrylate) (7.98 wt %) to modify the compatibility between fibre and PLA matrix.

Due to the large fraction of cellulose in the rice/wheat husks, they have potential to be utilized in biocomposite as reinforcement. In the study of García et al. [43], the rice husk-PLA composites did not show significant improvement in mechanical properties, compare to the pure PLA matrix. However, Tran el al. [65] evaluated the properties of the husk-PLA composites that underwent different fibre treatments, They found that the combination of an alkaline and an silane treatments could provide higher mechanical properties and thermal stability of the husk- PLA composite compared to the composite with only silane treatment.

Another abundant waste material from agriculture product is corncob. Similar to straw/grass fibre, corncob has potential as a filler in a biocomposite due to the low price and high availability. Although the addition of corncob in a PLA composite could offer enhancement in tensile modulus, the tensile strength and elongation at break of the corncob-PLA composite are lower than those of the PLA matrix because of the poor interfacial adhesion [66]. However, the tensile properties of corncob-PLA composites can be improved by using a coconut oil coupling agent [66].

Seed and Fruit Fibres Reinforced PLA Composites

Coconut coir fibre is extracted from coconut husk with hemicellulos and lignin [67, 68]. It has potential applications as reinforcement of a biodegradable composite in automobile and construction due to its low thermal conductivity and bulk density, as well as the low cost. The coir fibre also has higher elongation at break, but lower tensile strength and modulus, compared to other natural fibres because it contains low cellulose and hemicellulose [67, 68]. Short coir fibre reinforced PLA biocomposites fabricated by injection moulding show significantly higher tensile modulus than neat PLA matrix, and the tensile modulus increases with the increase of fibre mass content within the range of 10- 40% [68]. Dong et al. [67] found that short coir fibre reinforced PLA composites made by hot compression had lower tensile and flexural properties than neat PLA. However, alkali treated coir fibre composites had enhanced mechanical properties compared to the untreated fibre composites [67, 68]. This could be due to the alkali treatment improving fibre wetting and interfacial adhesion between PLA polymer and coir fibre [67, 68].

A cotton fibre reinforced PLA composite was developed by Graupner et al. [28]. The mechanical properties of the neat PLA are slightly improved by the addition of cotton fibre into the PLA matrix. The cotton-PLA composite shows lower tensile strength and modulus, but higher elongation at break and impact strength compared with hemp- and kenaf-PLA composites. Cotton linter, normally a waste material, is a mixture of residual cotton lint and cotton linters left on the cotton seed after ginning [55] . It was also the third largest fibre source after wood and bamboo in 2001 [69]. Way and the co-workers [55] developed a cotton linter-PLA composites by extrusion followed by injection moulding, and evaluated their mechanical properties and biodegradability. The tensile and flexural moduli and impact strength of the cotton linter –PLA composite with 25 wt% fibre content are 50 - 70% higher than those of the PLA matrix.

Biodegradability of Natural Fibre Reinforced PLA Composites

The biodegradability of PLA is highly dependent on the exposure environment. The conditions of the biodegradation environment include the presence of PLA degrading microorganisms, temperature, humidity and pH value. Many studies have evaluated the biodegradability of PLA based composites and compared it to that of neat PLA. The findings are varied, likely due to the varying test conditions. The PLA composites tend to be degraded faster at higher temperature (58 - 80°C) and in a enzymatic environment, compared to the soil burial test at room temperature [39, 55, 70, 71]. Most of the natural fibre reinforced PLA composites show faster biodegradability by burial test than the neat PLA matrix [31, 36, 47, 54, 55, 67, 70, 71]. Several studies also reported that although the mechanical properties of natural fibre reinforced PLA composite can be enhanced by fibre surface treatment, the biodegradability of the composites is adversely affected by the treatment [47, 54]. The addition of these chemicals may also bring about harmful chemical residues in the composting environment at the end-lift-cycle.

PLA Self-Reinforced Composites in Biomedical Application

The concept of self-reinforcement polymer composites (SRC) was first discovered by Capiti and Poter [72] to solve the interfacial adhesion problem in traditional composites. In their study, the self-reinforcement or all polymer composite is defined as the composite in which the matrix and reinforcement are made from similar or identical polymer but in different morphologies. Nowadays, the concept of SRC is widely used for thermoplastic polymer composite based on polyethylene, poly (propylene) and poly (ethylene terephthalate) polymers.

Self-reinforced bioabsorbable composite implants have been broadly used in tissue engineering, especially implants for bone tissue fixation applications [73, 74]. The implants are required to retain the tissue supporting properties for a given period and degrade gradually [73]. PLA, due to its high initial modulus as well as excellent biodegradability and biocompatibility, has been successfully used as the implants, such as screws and scaffolds for bone fracture fixation [74-77].

A patent [78] first described the surgical osteosynthesis composite material which consisted of the absorbable polymer or copolymer as matrix and reinforcement with the same chemical element. The bioabsorbable self-reinforcing composites can be produced by either extrusion method [73-76, 79] or hot compaction method [77, 79-82]. In the former method, a self-reinforced structure is induced into the matrix by transferring part of the microstructure of the PLA polymer into oriented elements such as macro/micro-fibrils, extended-chain crystals via mechanical deformation processes, such as free drawing, die drawing, shearing or rolling [79]. The PLA self-reinforced composites with good mechanical modulus, strength and toughness have been successfully produced by extrusion followed by die-drawing for bone fracture fixation applications [73-76, 79].

The hot compaction method is based on partial melting or softening of the PLA polymer fibres in a pressurized mould, and coalescing the melted fibre surfaces together during cooling [78, 79]. This method usually requires precise control of the heating temperature and pressure to give a thin melted/softened surface of fibres and to avoid destruction of reinforcement by over-heating. Charles et al. [80] found that the flexural modulus is significantly affected by the compaction temperature and time in the preparation of PLLA SRCs at fixed pressure (50 MPa) , and increases with the increase of compaction temperature.

In Nazhat et al. 's study [82], PLLA fibre and/or hydroxyapatite particle were reinforced with PDLLA film to create a biodegradable SRC for bone repairing and augmentation by hot compression. The compression moulding temperature of 100°C is lower than the onset melting temperatures of both the PDLLA matrix (122.8°C) and the PLLA fibre (167.2°C). A good interfacial bonding between the two phases may not be achieved at this processing temperature, but the use of a high pressure of 57 MPa in the study helped to build up a strong interfacial bonding.

New Developments in PLA Based Biocomposites

New Research on Bio-Fibres Reinforced PLA Composites

The reinforcements in the PLA based composites are not limited to natural fibres such as bast fibres and leaf fibres. In recent years, research on waste bio-fibres reinforced with the biodegradable polymer, PLA, has attracted wide-spread attention because these waste bio-fibres are low cost and abundantly available.

Recycled newspapers are becoming a more valuable source of fibre all over the world. The reuse of recycled newspapers includes cardboard, boxes, pencil barrels, tissue paper and other derivatives. Recycled newspapers are usually composed of lignocellulose and other inorganic materials such as printing ink [27]. Huda et al. [27] found the tensile and flexural moduli of PLA are significantly increased by adding 30% recycled newspaper cellulose fibre. The improvements of tensile and flexural moduli are 96% and 63%, respectively.

A large amount of chicken feather, about 2×10^9 kg per year in the United States, is discarded from the poultry sector and disposed by burial [83]. Chicken feather fibre contains hydrophobic keratin, which has similar strength as nylon fibre and smaller diameter than the wood fibre [83]. Chicken feather fibre is also a semi-crystalline material with cross-lined structure. Thus, its elastic modulus is comparable to coir fibre, approximately 3.4-5 GPa [83]. A chicken feather fibre reinforced PLA composite was developed by extrusion and injection moulding by Cheng et al. [83]. The tensile modulus of the chicken feather fibre (5 wt%)-PLA composite (4.2 GPa) is significantly higher than that of the neat PLA (3.6 GPa); there is no obvious change in tensile strength by adding chicken feather fibre into the PLA matrix. The chicken feather fibre composite also shows a ductile tensile behaviour. This indicates that the composite could provide an extended fracture process to avoid sudden failure [83].

The marine plant natural fibre, such as red algae fibre, has been used as reinforcement in PLA composite because the bleached red algae fibre (BRAF) has more uniform shape than other natural fibres and high thermal stability [84]. The BRAF-PLA composites with various fibre contents were produced by melt blending in the study of Sim et al., [84]. The dynamic mechanical and thermo-mechanical properties of the BRAF-PLA are significantly higher than those of pure PLA and increase with the fibre content up to 50%. Compare to the BRAF-PP composite, BRAF-PLA has lower coefficient of thermal expansion. It indicates that the PLA composite has higher dimensional stability, which could result in better interfacial adhesion between BRAF and PLA [84].

Okra bahmia is abundantly available in Indian, Mediterranean and Brazil. Instead of disposing of the okra as wasted material, it is possible to use it as a filler in biocomposites. Okra fibre is extracted from okra bark and can be utilized in the food sector, such as an added nutrient in gum and pectin [71]. Fortunati et al. [71] developed treated and untreated okra fibre reinforced PLA composites with different fibre contents (10-30 wt%). They found that all the okra-PLA composites have higher tensile modulus than the neat PLA. However, the fibre treatment did not provide enhancement in tensile modulus and stress at yield and break compared to the untreated okra-PLA composites with the same fibre contents.

PLA Single Polymer and PLA Fibre Reinforced Composites

Recently, the use of PLA polymer in the composite industry has been increasing rapidly due to its sustainability and biodegradability. However, PLA production usually requires agricultural products and the growth of PLA production could lead to great strain in food supply. The recycle of PLA is becoming more important. Research focus of PLA self-reinforced composite (SRC) has expanded from biomedical materials to the application for packaging and the other high-performance materials. The PLA SRC can be produced by numerous manufacturing techniques.

Li and Yao used the principle of nucleation of slowly crystallizing polymer to produce PLA SRCs in which amorphous PLA sheets was the matrix and a layer of PLA fabric was the reinforcement [85]. When these amorphous polymers are heated from room temperature, the glass-transition and crystallization processes occur simultaneously. The glass-transition (fusion) process will make the amorphous material soft and rubbery; whereas the crystallization process will turn it into a hardened phase. Thus, a short holding time is required to soften the matrix but minimise the crystallization. As shown in Figure 4, T_{min} is the minimum processing temperature, where the fusion time is significantly shorter than the crystallization time. Thus, the fusion process starts to dominate over the crystallization process from T_{min} [85, 86]. By using the method of slowly crystallizing polymer nucleation, a large processing temperature range (50 - 100ºC) can be obtained [85, 86]. However, sufficient matrix wetting may not be achieved at low processing temperatures. In Li and Yao's study, a wide processing temperature range from 100 - 150°C was used, although only temperatures over 130° C achieved good bonding between PLA fibre and matrix. The PLA SRCs exhibit significant improvements in mechanical properties compared to the unreinforced PLA sheet.

In the study of Jia el al., PLA SRCs with different fibre contents were fabricated via film-stacking followed by hot pressing [87, 88]. The processing temperature of 150°C is 10°C higher than the onset melting temperature of the amorphous PLA film matrix but lower than that of the crystalline PLA fibre. Due to the good interfacial bonding achieved by sufficient PLA matrix melting, the tensile modulus of the composites increases with the increase of fibre content in the loading direction and is up to 2 times higher than the neat PLA film. A PLA fibre reinforced polybutylene succinate (PBS) composite was also produced by the same method but lower processing temperature (130°C), owing to the low melting temperature of the PBS polymer. Compared to the PLA SRCs, the PLA-PBS composites with the same fibre contents have lower tensile modulus and strength, but higher strain at break [87, 88]. This is because the PLA SRC has stronger interfacial bonding than the PLA–PBS composite, as shown in Figure 5, the pull-out fibres from the PBS matrix (Figure 5a) are much longer than those from the PLA matrix (Figure 5b).

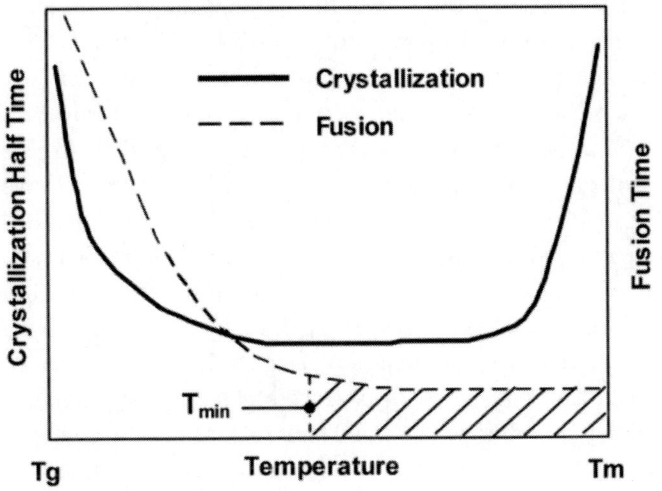

Figure 4. Feasible process window for SRC by nucleation method [86].

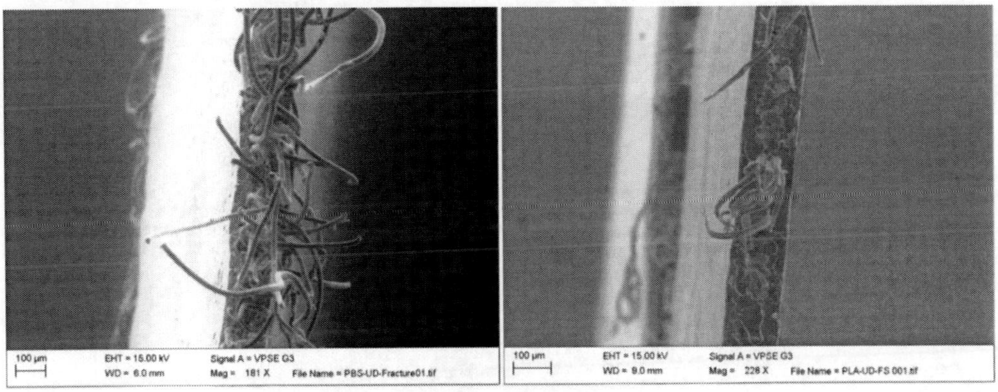

Figure 5. Tensile fracture surfaces of PLA-PBS composite (a) and PLA SRC (b) [87, 88].

Three-Dimensional PLA Biocomposite

Three-dimensional (3D) composite is characterized by high mechanical properties in multi-directions including the through thickness direction. A 3D composite usually has high impact resistance and delamination resistance. In recent years, 3D biocomposites have attracted intensive research interest. 3D woven structures, such as angle-interlock and orthogonal, play an important role in the 3D composite industry. In the study of Zhou et al. [89], orthogonal woven structured ramie-PLA composite preforms were fabricated by co-wrapped yarns which consist of ramie staple yarn as core yarn and PLA yarn as binding yarn. During the hot pressing process, the PLA yarn melts and bonds with the ramie yarn to form the composite. For 3D thermoplastic composites, the poor resin impregnation is a common problem. However, by using co-wrapped yarn, a good matrix penetration can be achieved in the orthogonal structured composite, despite textile preforms have a high thickness of 3.2-3.7 mm.

In Wu et al.' study, 3D and 5-directional braiding technique were also used to fabricate PLA textile composite preforms by using different braiding angles [90]. During the hot compaction process, the PLA fibres of the braided fabric were partially melted to fuse together. The flexural modulus and strength were increased with the increase of consolidation temperature from 130°C to 150°C because the bonding degree of PLA fibres was enhanced; they also increase as the preform thickness increase from 5 to 8 mm. However, the increase of braiding angle only provides significant improvement on flexural strength.

Nonwoven technique is also popularly used in making 3D composites because nonwoven structures can provide a good inherent z-directional property [91]. In Alimuzzaman et al.'s study, a 3D flax/PLA nonwoven shell was firstly developed by using a web-forming system with a bonding chamber, and with the help of a designed hemisphere shaped mould, a dome-shaped flax reinforced PLA composite was produced by hot compression moulding, as shown in Figure 6 [91]. They found that, as the flax fibre content increased, the crush failure load, total energy absorption, specific energy absorption and crush efficiency increased. The highest crushing property was achieved by the 3D flax-PLA composite with 60% fibre content and made at 180 °C and 5 min moulding temperature and time.

Figure 6. Dome-shaped flax-PLA biocomposite [91].

Conclusion

PLA based biocomposites have attracted intensive interest in both academic and industrial fields because the composites can reduce the negative impact on the environment. Natural fibre reinforced PLA composites, such as flax-, hemp- and sisal-PLA composites show equal or even better mechanical properties compared to glass fibre reinforced composite. They have been successfully used in automotive parts (e.g. doors, trunks and floor panels) and construction materials (e.g. ceiling, floor and window). New research on the enhancements of the properties of natural fibre reinforced PLA composites is being carried out in order to discover broader applications including furniture, suitcase, electronic products, etc. The fibres that are extracted from wasted agriculture plants have potential as fillers in PLA composites. However, PLA is still considered as a high-priced material and more production of PLA may result in strains in food supply. Therefore, the improvements in recyclability and utilization of PLA are becoming more important. PLA self-reinforced composite and the separation of PLA from natural fibre-PLA composite are important future research topics.

References

[1] Jamshidian, M., et al., Poly-Lactic Acid: Production, Applications, Nanocomposites, and Release Studies. *Comprehensive Reviews in Food Science and Food Safety,* 2010. 9(5): p. 552-571.

[2] Balkcom, M., B. Welt, and K. Berger, Notes from the packaging labortory: PLA-- An exciting new packaging material, 2002, The University of Florida: Florida.

[3] Blackburn, R.S., Introduction, in Biodegradable and sustainable fibres, R.S. Blackburn, Editor. 2005, Woodhead publishing limited: Cambridage.

[4] Farrington, D.W., et al., Poly(lactic acid) fibres, in Biodegradable and sustainable fibres, R.S. Blackburn, Editor. 2005, Woodhead publishing limited: Cambridae. p. 191-220.

[5] Henton, D., et al., Polylactic Acid Technology, in *Natural fibres, biopolymer, and biocomposites.* 2005. p. 527-579.

[6] Moon, S.I., et al., Melt/solid polycondensation of l-lactic acid: an alternative route to poly(l-lactic acid) with high molecular weight. *Polymer,* 2001. 42: p. 5059-5062.

[7] Avinc, O.O., Studies of the effects of processing conditions and softeners on the physical characteristics and performance of poly(lactic acid)- based fabrics, in School of Materials 2007, The University of Manchester: Manchester.

[8] Ikada, Y., et al., Stereocomplex formation between enantiomeric poly(lactides). *Macromolecules,* 1987. 20(4).

[9] Perego, G. and G.D. Cella, Mechanical properties, in POLY(LACTIC ACID)-Synthesis, Structures, Properties, Processing, and Applications, R. Auras, et al., Editors. 2010, *John Wiley & Sons, Inc.*: New Jersey.

[10] Perego, G., G.D. Cella, and C. Bastioli, Effect of molecular weight and crystallinity on poly (lactic acid) mechanical properties. *Journal of applied polymer science,* 1996. 59(1): p. 37-43.

[11] Agrawal, A.K., Spinning of poly(lactic acid) fibres, in POLY(LACTIC ACID)-Synthesis, Structures, Properties, Processing, and Applications, R. Auras, et al., Editors. 2010, *John Wiley & Sons, Inc.:* New Jersey.

[12] Solarski, S., M. Ferreira, and E. Devaux, Characterization of the thermal properties of PLA fibers by modulated differential scanning calorimetry. *Polymer,* 2005. 46(25): p. 11187-11192.

[13] Calabia, B.P., et al., Biodegradation, in POLY(LACTIC ACID)-Synthesis, Structures, Properties, Processing, and Applications, R. Auras, et al., Editors. 2010, *John Wiley & Sons,* Inc.: New Jersey.

[14] Garlotta, D., A literature rewiew of poly(lactic acid). *Journal of polymers and the enviroment,* 2002. 9(2): p. 63-84.

[15] Meinander, K., et al., Polylactides-degaradable polymers for fibres and films. *Macromol. Symp.,* 1997. 123: p. 147-153.

[16] Hull, D. and T.W. Clyne, An introduction to composite materials. 2nd ed. Cambridage Solid State Science Series. 1996, New York: *Cambridge University Press.*

[17] Weston, H., Principles of adhesion and bonding in Flexible textile composites P. Lennox-Kerr., Editor. 1973, *The Textile Trade Press: Manchester.*

[18] Drzal, L.T., A. Mohanty, and M. Misra, Bio-composite materials as alternatives to petroleum-based composites for automotive applications. *Magnesium,* 2001. 40(60): p. 1.3-2.

[19] Mohanty, A., M. Misra, and L. Drzal, Sustainable bio-composites from renewable resources: opportunities and challenges in the green materials world. *Journal of Polymers and the Environment,* 2002. 10(1-2): p. 19-26.

[20] Mohanty, A., M. Misra, and G. Hinrichsen, Biofibres, biodegradable polymers and biocomposites: an overview. *Macromolecular Materials and Engineering,* 2000. 276(1): p. 1-24.

[21] Koronis, G., A. Silva, and M. Fontul, Green composites: A review of adequate materials for automotive applications. *Composites Part B: Engineering*, 2012.

[22] Bajpai, P.K., I. Singh, and J. Madaan, Development and characterization of PLA-based green composites: A review. *Journal of Thermoplastic Composite Materials*, 2012.

[23] Ghosh, S.B., S. Bandyopadhyay-Ghosh, and M. Sain, Composites in POLY(LACTIC ACID)-Synthesis, Structures, Properties, Processing, and Applications, R. Auras, et al., Editors. 2010, *John Wiley & Sons,* Inc.: New Jersey.

[24] Bühler, M., P.-E. Bourban, and J.-A.E. Månson, Cellular composites based on continuous fibres and bioresorbable polymers. *Composites Part A: Applied Science and Manufacturing,* 2008. 39(12): p. 1779-1786.

[25] Ahmed, I., et al., Retention of mechanical properties and cytocompatibility of a phosphate-based glass fiber/polylactic acid composite. *Journal of Biomedical Materials Research Part B: Applied Biomaterials,* 2009. 89B(1): p. 18-27.

[26] Ahmed, I., et al., Composites for bone repair: phosphate glass fibre reinforced PLA with varying fibre architecture. *Journal of materials science: materials in medicine,* 2011. 22(8): p. 1825-1834.

[27] Huda, M.S., et al., Chopped glass and recycled newspaper as reinforcement fibers in injection molded poly (lactic acid)(PLA) composites: a comparative study. *Composites science and technology,* 2006. 66(11): p. 1813-1824.

[28] Graupner, N., A.S. Herrmann, and J. Mussig, Natural and man-made cellulose fibre-reinforced poly(lactic acid) (PLA) composites: An overview about mechanical characteristics and application areas. *Composite: Part A,* 2009. 40: p. 810-821.

[29] Bax, B. and J. Müssig, Impact and tensile properties of PLA/Cordenka and PLA/flax composites. *Composites science and technology,* 2008. 68(7): p. 1601-1607.

[30] Ganster, J. and H.-P. Fink, Novel cellulose fibre reinforced thermoplastic materials. *Cellulose,* 2006. 13(3): p. 271-280.

[31] Islam, M.S., K.L. Pickering, and N.J. Foreman, Influence of accelerated ageing on the physico-mechanical properties of alkali-treated industrial hemp fibre reinforced poly

(lactic acid)(PLA) composites. *Polymer Degradation and Stability,* 2010. 95(1): p. 59-65.

[32] Hu, R. and J.K. Lim, Fabrication and mechanical properties of completely biodegradable hemp fiber reinforced polylactic acid composites. *Journal of composite materials,* 2007. 41(13): p. 1655-1669.

[33] Sawpan, M.A., K.L. Pickering, and A. Fernyhough, Improvement of mechanical performance of industrial hemp fibre reinforced polylactide biocomposites. *Composites Part A: Applied Science and Manufacturing*, 2011. 42(3): p. 310-319.

[34] Plackett, D., et al., Biodegradable composites based on L-polylactide and jute fibres. *Composites science and technology,* 2003. 63: p. 1287-1296.

[35] Khondker, O.A., et al., A novel processing technique for thermoplastic manufacturing of unidirectional composites reinforced with jute yarns. *Composites Part A: applied science and manufacturing,* 2006. 37: p. 2274-2284.

[36] Alimuzzaman, S., R. Gong, and M. Akonda, Biodegradability of nonwoven flax fiber reinforced polylactic acid biocomposites. *Polymer Composites,* 2014.

[37] Zhang, W. and Y. Li, Fabrication and characterization of flax fiber reinforced polylactic acid composite [J]. *Journal of Textile Research,* 2009. 6: p. 010.

[38] Alimuzzaman, S., R.H. Gong, and M. Akonda, Nonwoven polylactic acid and flax biocomposites. *Polymer Composites,* 2013. 34(10): p. 1611-1619.

[39] Ochi, S., Mechanical properties of kenaf fibers and kenaf/PLA composites. *Mechanics of Materials,* 2008. 40(4–5): p. 446-452.

[40] Huda, M.S., et al., Effect of fiber surface-treatments on the properties of laminated biocomposites from poly (lactic acid)(PLA) and kenaf fibers. *Composites science and technology,* 2008. 68(2): p. 424-432.

[41] Ben, G., et al., Examination of heat resistant tensile properties and molding conditions of green composites composed of kenaf fibers and PLA resin. *Advanced Composite Materials,* 2007. 16(4): p. 361-376.

[42] Serizawa, S., K. Inoue, and M. Iji, Kenaf-fiber-reinforced poly (lactic acid) used for electronic products. *Journal of applied polymer science,* 2006. 100(1): p. 618-624.

[43] García, M., I. Garmendia, and J. Garcia, Influence of natural fiber type in eco-composites. *Journal of applied polymer science,* 2008. 107(5): p. 2994-3004.

[44] Oksman, K., M. Skrifvars, and J.-F. Selin, Natural fibres as reinforcement in polylactic acid (PLA) composites. *Composites science and technology,* 2003. 63: p. 1317-1324.

[45] Sawpan, M., K. Pickering, and A. Fernyhough, Hemp fibre reinforced poly (lactic acid) composites. *Advanced Materials Research,* 2007. 29: p. 337-340.

[46] Prajer, M. and M.P. Ansell, Bio-composites for structural applications: Poly-l-lactide reinforced with long sisal fiber bundles. *Journal of applied polymer science,* 2014: p. n/a-n/a.

[47] Wu, C.-S., Preparation, characterization, and biodegradability of renewable resource-based composites from recycled polylactide bioplastic and sisal fibers. Journal of applied polymer science, 2012. 123(1): p. 347-355.

[48] Zhang, M., et al., Effect of transcrystallinity on tensile behaviour of discontinuous carbon fibre reinforced semicrystalline thermoplastic composites. *Polymer,* 1996. 37(23): p. 5151-5158.

[49] Teishev, A. and G. Marom, The effect of transcrystallinity on the transverse mechanical properties of single-polymer polyethylene composites. *Journal of applied polymer science,* 1995. 56(8): p. 959-966.

[50] Wang, Y., et al., Transcrystallization behavior at the poly (lactic acid)/sisal fibre biocomposite interface. *Composites Part A: Applied Science and Manufacturing,* 2011. 42(1): p. 66-74.

[51] Bajpai, P.K., I. Singh, and J. Madaan, Tribological behaviour of natural fiber reinforced PLA composites. *Wear,* 2012.

[52] Shih, Y.-F. and C.-C. Huang, Polylactic acid (PLA)/banana fiber (BF) biodegradable green composites. *Journal of Polymer Research*, 2011. 18(6): p. 2335-2340.

[53] Asaithambi, B., G. Ganesan, and S.A. Kumar, Bio-composites: Development and mechanical characterization of banana/sisal fibre reinforced poly lactic acid (PLA) hybrid composites. *Fibers and Polymers,* 2014. 15(4): p. 847-854.

[54] Shibata, M., et al., Biocomposites made from short abaca fiber and biodegradable polyesters. *Macromolecular Materials and Engineering,* 2003. 288(1): p. 35-43.

[55] Way, C., et al., Processing stability and biodegradation of polylactic acid (PLA) composites reinforced with cotton linters or maple hardwood fibres. *Journal of Polymers and the Environment,* 2013. 21(1): p. 54-70.

[56] Pilla, S., et al., Polylactide-pine wood flour composites. *Polymer Engineering & Science,* 2008. 48(3): p. 578-587.

[57] Hrabalova, M., et al., Effect of wood flour loading and thermal annealing on viscoelastic properties of poly (lactic acid) composite films. *Journal of applied polymer science,* 2010. 118(3): p. 1534-1540.

[58] Du, Y., et al., Fabrication and characterization of fully biodegradable natural fiber-reinforced poly(lactic acid) composites. *Composites Part B: Engineering,* 2014. 56(0): p. 717-723.

[59] Ma, H. and C.W. Joo, Influence of surface treatments on structural and mechanical properties of bamboo fiber-reinforced poly (lactic acid) biocomposites. *Journal of composite materials,* 2011: p. 0021998311401096.

[60] Tokoro, R., et al., How to improve mechanical properties of polylactic acid with bamboo fibers. *Journal of Materials Science,* 2008. 43(2): p. 775-787.

[61] Lee, S.-H., T. Ohkita, and K. Kitagawa, Eco-composite from poly (lactic acid) and bamboo fiber. *Holzforschung,* 2004. 58(5): p. 529-536.

[62] Rawi, N.F.M., K. Jayaraman, and D. Bhattacharyya, A performance study on composites made from bamboo fabric and poly (lactic acid). *Journal of Reinforced Plastics and Composites,* 2013. 32(20): p. 1513-1525.

[63] Rawi, N.F.M., K. Jayaraman, and D. Bhattacharyya, Bamboo fabric reinforced polypropylene and poly(lactic acid) for packaging applications: Impact, thermal, and physical properties. *Polymer Composites,* 2013: p. n/a-n/a.

[64] Qin, L., et al., Mechanical and thermal properties of poly(lactic acid) composites with rice straw fiber modified by poly(butyl acrylate). *Chemical Engineering Journal,* 2011. 166(2): p. 772-778.

[65] Tran, T.P.T., J.-C. Bénézet, and A. Bergeret, Rice and Einkorn wheat husks reinforced poly(lactic acid) (PLA) biocomposites: Effects of alkaline and silane surface treatments of husks. *Industrial Crops and Products*, 2014. 58(0): p. 111-124.

[66] Chun, K.S. and S. Husseinsyah, Polylactic acid/corn cob eco-composites: Effect of new organic coupling agent. *Journal of Thermoplastic Composite Materials,* 2013: p. 0892705712475008.

[67] Dong, Y., et al., Polylactic acid (PLA) biocomposites reinforced with coir fibres: Evaluation of mechanical performance and multifunctional properties. *Composites Part A: Applied Science and Manufacturing,* 2014. 63(0): p. 76-84.

[68] Nam, T.H., S. Ogihara, and S. Kobayashi, Interfacial, Mechanical and Thermal Properties of Coir Fiber-Reinforced Poly (Lactic Acid) *Biodegradable Composites. Advanced Composite Materials,* 2012. 21(1): p. 103-122.

[69] Eichhorn, S.J., et al., Review: Current international research into cellulosic fibres and composites. *Journal of Materials Science,* 2001. 36(9): p. 2107-2131.

[70] Ohkita, T. and S.H. Lee, Thermal degradation and biodegradability of poly (lactic acid)/corn starch biocomposites. *Journal of applied polymer science,* 2006. 100(4): p. 3009-3017.

[71] Fortunati, E., et al., Okra (Abelmoschus Esculentus) fibre based PLA composites: mechanical behaviour and biodegradation. *Journal of Polymers and the Environment,* 2013. 21(3): p. 726-737.

[72] Capiati, N.J. and R.S. Porter, The concept of one polymer composites modelled with high density polyethylene. *Journal of materials science*, 1975. 10: p. 1671-1677.

[73] Niemelä, T., et al., Self-reinforced composites of bioabsorbable polymer and bioactive glass with different bioactive glass contents. Part I: Initial mechanical properties and bioactivity. *Acta biomaterialia,* 2005. 1(2): p. 235-242.

[74] Kellomäki, M., et al., Bioabsorbable scaffolds for guided bone regeneration and generation. *Biomaterials,* 2000. 21(24): p. 2495-2505.

[75] Pohjonen, T., et al., Strength retention of self-reinforced poly-L-lactide screws. A comparison of compression moulded and machine cut screws. *Journal of materials science: materials in medicine,* 1997. 8(5): p. 311-320.

[76] Saikku-BÄCkstrÖM, A., et al., Material properties of absorbable self-reinforced fibrillated poly-96L/4 D-lactide (SR-PLA96) rods; a study in vitro and in vivo. *Journal of materials science: materials in medicine,* 1999. 10(1): p. 1-8.

[77] Majola, A., et al., Absorbable self-reinforced polylactide (SR-PLA) composite rods for fracture fixation: strength and strength retention in the bone and subcutaneous tissue of rabbits. *Journal of materials science: materials in medicine* 1992. 3: p. 43-47.

[78] Törmälä, P., et al., Material for osteosynthesis devices, in United States Patent1988, Materials consultants: United States.

[79] Tormala, P., Biodegradable self-reinforced composite materials; manufacturing structure and mechanical properties. *Clinical Materials*, 1992. 10: p. 29-34.

[80] Charles, L.F., et al., Self-reinforced composites of hydroxyapatite-coated PLLA fibers: fabrication and mechanical characterization. *Journal of the mechanical behavior of biomedical materials,* 2013. 17: p. 269-277.

[81] Wright-Charlesworth, D.D., et al., Nanoindentation of injection molded PLA and self-reinforced composite PLA after in vitro conditioning for three months. *Journal of Biomedical Materials Research Part A,* 2005. 74(3): p. 388-396.

[82] Nazhat, S.N., et al., Dynamic mechanical characterization of biodegradable composites of hydroxyapatite and polylactides. *J Biomed Mater Res,* 2001. 58: p. 335-343.

[83] Cheng, S., et al., Mechanical and thermal properties of chicken feather fiber/PLA green composites. *Composites Part B: Engineering,* 2009. 40(7): p. 650-654.

[84] Sim, K.J., S.O. Han, and Y.B. Seo, Dynamic mechanical and thermal properties of red algae fiber reinforced poly (lactic acid) biocomposites. *Macromolecular Research,* 2010. 18(5): p. 489-495.

[85] Li, R. and D. Yao, Preparation of single poly(lactic acid) composites. *Journal of applied polymer science,* 2008. 107(5): p. 2909-2916.

[86] Yao, D., R. Li, and P. Nagarajan, Single-polymer composites based on slowly crystallizing polymers. *Polymer Engineering & Science,* 2006. 46(9): p. 1223-1230.

[87] Jia, W., et al., Biodegradable fibre reinforced composites composed of polylactic acid and polybutylene succinate. *Plastics, Rubber and Composites,* 2014. 43(3): p. 82-88.

[88] Jia, W., R.H. Gong, and P.J. Hogg, Poly (lactic acid) fibre reinforced biodegradable composites. Composites Part B: *Engineering,* 2014. 62: p. 104-112.

[89] Zhou, N., et al., Mechanical and sound adsorption properties of cellular poly (lactic acid) matrix composites reinforced with 3D ramie fabrics woven with co-wrapped yarns. *Industrial Crops and Products,* 2014. 56(0): p. 1-8.

[90] Wu, N., et al., Preparation and bending properties of three dimensional braided single poly (lactic acid) composite. *Composites Part B: Engineering,* 2013. 52: p. 106-113.

[91] Alimuzzaman, S., R.H. Gong, and M. Akonda, Three-dimensional nonwoven flax fiber reinforced polylactic acid biocomposites. *Polymer Composites*, 2014. 35(7): p. 1244-1252.

In: New Developments in Polylactic Acid Research
Editor: Courtney Winthrop
ISBN: 978-1-63463-054-2

Chapter 5

Applications of Polylactic Acid in Surgery of Abdominal Wall Hernias

***Juan Manuel Suárez Grau, MD, PhD*[1], *Cristina Méndez García, MD*[2], *Salud García Ruiz, MD*[3], *Carmen Cepeda Franco, MD*[4], *Carolina Rubio Chaves, NUR*[4] *and Fernando Docobo Durántez, MD, PhD, FACS*[4]**

[1]General Hospital of Riotinto, Huelva, Spain
Clinic Quirón "Sagrado Corazón", Seville, Spain
[2]General Hospital of Jerez, Cádiz, Spain
[3]Clinic Quirón "Donostia", Guipúzcoa, Spain
[4]University Hospital Virgen del Rocío, Seville, Spain

Abstract

1) Use in Sutures: a) Absorbables sutures with polylactic. Polyglycolic acid, used since 1968, is obtained by polymerization of glycolic acid, the acid which exists in the sugar cane. It is a synthetic product, it is absorbed on 9 months at the most by chemical hydrolysis. It causes minimal tissue reaction and its tensile strength is superior to catgut, silk and cotton. Some derivates are used in Surgery: Polydioxanone acid, glycolic Poly caprolactone, Lactomer Copolymer, Polyglecaprone 25, Polyglactin 910.

b) Self-adhesions sutures. A bearded self-fixing system allowing the suture material for tissue approximation, which eliminates a long and tedious step in wound closure: the knoting to attach and secure the connection. This new system provides a more uniform tension on the soft tissue as they approach so stress is distributed over the entire wound lenght; thus, placement of the sutures with self-locking is less traumatic to tissue as it eliminates the need for sutures or discontinued knots.

c) Active sutures. The new active Plus Vicryl (polyglactin 910 antibacterial coated with triclosan) has been proved as an option for closing wounds, especially in those with active infection or septic environment.

This suture has more effective antibacterial ability, very useful in the prevention of postsurgical infection even using prosthetic material.

Vicryl Plus, the first antibacterial suture was a novel antibacterial designed synthetic suture preventing bacterial colonization in the suture. Vicryl Plus antibacterial agent incorporates Irgacare®, a broad-spectrum antiseptic in form pure triclosan and is known for its effectiveness against bacteria that cause most infections surgical, among which are included Staphylococcus aureus, S. epidermidis sensitive and resistant to methicillin.

Vicryl Plus, to be coated with triclosan, generates a zone of inhibition around the suture and decreases the risk of infection by preventing colonization by bacteria. This eliminates one of the factors contribute to infection, because 66 % of surgical infections are located exclusively on wound.

2) Use in meshes: a) Hibrid synthetic meshes. *Lightweight prostheses:* these meshes have a big pore and low density. This prostheses *lightweight* can be made so types of material (one single material or mixed materials such non-absorbable and absorbable, such the polylactid acid).

These large pore and low density implants causes less strange body reaction, because have less biomaterial, with the advantage that the implant maintains the elasticity of the host tissue. Lightweight prostheses can be either non-absorbable material as absorbable material, thereby generating hybrid prostheses. These hybrid prostheses generates less fibrous reaction favoring the integration of the mesh and obtaining improved clinical outcomes compared with open surgery.

Composites prostheses: This prosthesis provokes high formation of peritoneum and high integration. The composite prosthesis consists in: Two different- materials layers. The first (superior layer) use to be prosthesis of reticular type, destined to increase the fibroblastic reaction, polypropylene or polyester usually. The second component (inferior layer) uses to be a laminar type and it can be absorbable or scarcely reactive material like the PTFE.

b) Absorbable Synthetic meshes. Up to the present, prostheses used for reconstruction surgery of the abdominal wall have been non-absorbable, which could be associated with chronic inflammatory processes causing complications in these meshes, such as chronic pain, infection, migration, intestinal obstruction, and fistulisation.

New degradable and absorbable materials may be the solution to these complications that, although infrequent, often become quite serious when they appear. These materials may also be applied to potentially contaminated areas. New synthetic absorbable prostheses have appeared recently such materials PGA-TMC (Bioabsorbable Plug Gore®), VIPRO (Vycryl, Ethicon®). These meshes have been used in various studies and they have shown decrease of postoperative pain, which is associated with smaller inflammatory process, being these materials degradable and absorbable.

Other proprieties of these meshes is the hydrophilic characterization, allowing the possibility of incorporates antiseptics or similar in water solution.

c) Absorbable grips in meshes (selfgripping- meshes). This is a partially absorbable mesh, low molecular weight polypropylene monofilament composed of polylactic acid (absorbable). The appendixes of polylactic acid, as a hook, are inserted between the muscle fibers, allowing secure attachment from its inner surface, with less risk of mobilization of the hernia sac under the prosthesis.

It is presented as the first true system of tension-free repair in the treatment of hernia pathology, with extensive coverage of the area being treated. The mesh Parietenne ® Progrip ® (Sofradim, Group Covidien, Trévoux, France) is presented in three modalities, two of them with an elliptical shape, specially designed to fit the anatomy of the inguinal region, with a self-adhesive flap adjustable cord around, either left or right. The third variant is a flat plate 15x9 cm, used for anterior abdominal wall.

The application of polylactic acid in surgery, has developed multiple advances as sutures and surgical mesh. Innovations are not only fit, also in coatings and connections with other products. The future is headed in abdominal wall surgery to mesh with absorbable polyglycolic acid derivatives in the short-medium term, with incorporation of biological products (stem cells, growth factors ...) and tridimensional disposition to facilitate the formation of new tissues the body itself.

1) Use in Sutures

a) Absorbables Sutures with Polylactic

Generally, sterile surgical sutures are used to close wounds, to ligate blood vessels, to maintain tissues together or to fix prosthetic implants. The materials used can be natural products, synthetic fibers and some polymer fibers specifically synthesized for use as suture materials. The last type is which takes our attention in this chapter. The evolution of synthetic suture material has reached a level of sophistication such that includes sutures designed for specific surgical procedures. Not just by eliminating some of the difficulties that the surgeon had found in wound closure but even decreasing the potential for post-operative infection. The wide variety of suture materials available on the market increases importance of understanding the basic properties to comply, in order to select the most appropriate for each situation material. The ideal suture does not yet exist, yet we know that the general properties could possess a good suture for each surgical procedure. These properties should include:

- High tensile strength and uniform, which allows the use of thin sizes
- Always smooth uniform diameter along the suture
- Flexibility for easy handling and knot security
- Absorbable Wound healed after
- Minimal tissue reaction
- Royalty irritants and impurities that promote growth bacterial
- Not having electrolyte, capillary (step liquid) or carcinogenic properties
- Sterile and ready for use
- Predictable behavior

As well no material has shown all these properties, some derivated of lactyc acid has been considered as a very close aproximation, because are well tolerated suture with high resistance and left no residuals when completely absorbed.

A first classification could in the monofilament sutures (single thread) and multifilament sutures (several strands or braided). Independiently of the material, monofilament sutures are better tolerated, have minimal tissue reaction, offer less resistance to its passage through the tissues, removal of points is less painful and leave minimal scarring; but instead, are difficult to handle, need more knots and are more expensive.

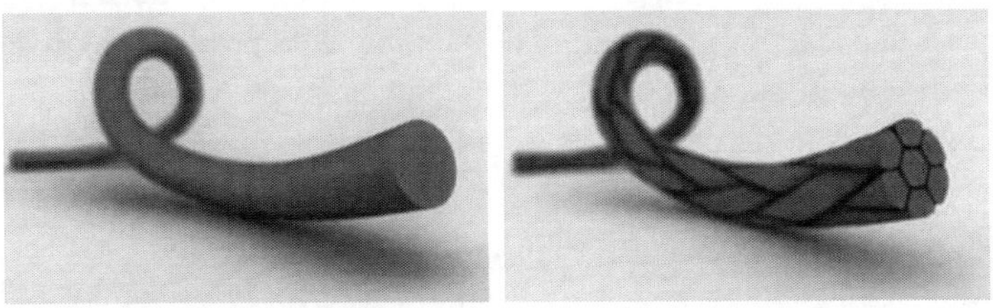

Diferences between monofilament and braided suture (multifilament).

Multifilament sutures have greater tissue reaction, so it should not be usedin infected wounds or if you suspect infection nor whether to keep long. Provide greater tensile strength , flexibility and easier handling, but leave more scarring. They also offer greater resistance to pass through the tissue, but to solve this problem are coated with waxes, silicones or fluorocarbons.

Synthetic derivates of poliglicolyc acid are hydrolyzed, that is, the water penetrates into the structure and dissolves. Obviously, the hydrolyzate of synthetic sutures is less aggressive to implementation of the enzyme system (in addition to attacking the suture will cause a number of injuries to the surrounding tissues).

Some of them are rapidly absorbed while others are treated or chemically structured to prolong absorption. They may also be coated or impregnated with agents which improve its handling properties and stained with dyes to increase their visibility in tissue.

The absorbable sutures, due to degradation suffering, decrease resistance over time. We must be clear that the tensile strength and absorption rate of the suture are separate elements. For example a suture can quickly lose its tensile strength in tissue but be absorbed very slowly; or it can retain its tensile strength during the time necessary for healing and then be absorbed quickly.

While offering many advantages, absorbable sutures also have certain limitations. If a patient has a fever or infection or suffer protein deficiency the absorption process can be accelerated. Also, if the sutures are wet or moistened during handling prior to use, the absorption may begin prematurely. All these characteristics depends on the specific polymer we are using.

Poli(ácido glicólico)	*Dexon, Safil*
Copolímero de PGA y PLA	*Vycril, Panacryl*
Poli(p-dioxanona)	*PDS, MonoPlus*
Copolímero de PGA y TMC	*Maxon*
Copolímero de PGA y PCL	*Monocryl*
Copolímero de PGA, PDO y TMC	*Biosyn*
Copolímero de PGA, PCL y TMC	*Monosyn*
Copolímero de PGA, PLA, PCL y TMC	*Caprosyn*

Equivalences between trademarks and manufacture.

Poliglycolic Acid (PGA)

Shaped suture braided multifilament composed of polyglycolic acid in a 99%, with caprolactone coated and calcium stearate. The coating will provides high tensile strength, excellent workability, low tissue reactivity, predictable absorption rate, and a firm and secure knotting. Ethylene oxide should be used as a method for sterilization is, because the gamma rays accelerate the lost of the tensile strength.

PGA sutures was a major step in suture materials absorbable, being physically and biologically superior to the catgut. Moreover, unlike collagen, PGA sutures exhibit no loss of mechanical strength to wetting, resulting in very little inflammatory response and have increased flexibility. After 15 days, the material has lost over 70% of its original strength. At 28 days, retained only 5%. It is absorbed completely passed between 70 and 100 days after implantation.

Copolymers of Poliglycolyc Acid and Lactic Acid (PGA-PLA)

Multifilament sutures consisting of a copolymer of glycolic acid and lactic. Currently on the market a wide range of such copolymers, being the most representative Polyglactin 910, sold under the tradename VICRYL, with an average composition of 90 % glycolic acid and 10 % lactic acid.

They are characterized by a loss of quite similar to poliglycolic sutures resistance: the material retains only 8% of its original strength after 28 days missing completely at 35 days after implantation. However, the time of complete absorption of the Vicryl sutures (50 to 70 days) is less than the Dexon, mainly due to the presence of the "lactide" groups in the main chain, which accelerates the hydrolysis of the material.

In some cases it is desirable to use coatings which improve basic material properties (Vycril plus). It is a mixture equally composed of polyglactin 370 (30 % glycolic acid and 70 % lactic acid), calcium stearate and stearic acid. This coating is absorbed rapidly and predictably at 90 days and inert.

Today another suture (PANACRYL) is marketed by a composite PGA-PLA copolymer, but the composition is just Vicryl reverse, with 10% glycolic acid and 90 % lactic acid. As coating uses a copolymer of e-caprolactone and glycolic acid.

Copolymers of Poliglycolyc Acid Y Polycaprolactone (PGA-PCL)

Monofilament suture composed by polycaprolactone block copolymers and polyglycolic acid, sold under the name MONOCRYL (poliglecaprone 25)

Has lower tensile strength than the suture Dexon since a week lost the 50%, after two weeks only maintains 25% of its initial tensile strength. It has excellent workability, low memory, low tissue reactivity and high elasticity. On average, takes between 90 and 120 days to completely dissolve.

It is the least reactive of all substance and less scar should leave. Unfortunately, it is a very soft material and can be slightly irritating when the stitches are not removed. Have also synthesized a copolymer of glycolide (72%), trimethylene carbonate (14%) and e-caprolactone (14%), to form a suture sold under the name of Monosyn. This absorbable suture is resistant, with good mechanical properties and easy to handle (due to its flexibility).

Polidioxanone (PDO)

Monofilament suture violet or colorless; violet color provides good visibility in the operative field and is unobtrusive colorless procedures for subdermal in plastic surgery. It is marketed under the names of PDS or MONOPLUS.

The poly (p-dioxanone) is a polyester-ether, which is different from poly (glycolic acid) because an ester group is replaced by an ether, more flexible, and also introduces a methylene group in the main chain with the resulting additional flexibility. This increased flexibility results in a lower reaction in host tissues.

Moreover, to decrease the concentration of ester groups is polydioxanone degrades more slowly than poly (a-hydroxy acid) (PGA and PLA), retaining his strength for longer periods of time. The original tensile strength is retains between 65 and 70 % in the first four weeks and between 50 and 60 % after the sixth week. Has a useful period of resistance between 35 and 40 days. Offers minimal tissue reactivity . Absorbed by slow progressive hydrolysis completing within 180 days about; absorption is minimal until after 90 days after implantation.

Copolymers of Glycolic Acid and Methylene Carbonate (PGA-TMC)

Monofilament suture born as an alternative to polydioxanone trying to improved flexibility of polyglycolic acid or polylactic acid copolymers thereof. It compounds is glycolic acid copolymers (PGA) and trimethylene carbonate (TMC). This product is currently available in the market with the name of MAXON.

This suture has a higher strength retention 59% after 28 days. 6 weeks after such sutures retain 30 % of its original strenght. Complete absorption takes place between 180 and 210 days with minimal tissue reaction. Have also synthesized a copolymer of glycolide (60 %), trimethylene carbonate (26 %) and polydioxanone (14 %), to form a suture sold under the name of BIOSYN. This suture retains about 75 % of its tensile strength at 14 days and 40% after three weeks. It is completely absorbed in 3-4 months and has low stiffness compared with pure PGA fibers.

There is another monofilament suture (CAPROSYN) synthesized from glycolic acidtrimethylene carbonate, lactic acid and polycaprolactone with excellent features manipulability. Maintains 50-60 % of its original strength after 5 days after implementation and 20-30% after 10 days; complete loss of tensile strength occurs at 21 some days. Is completely absorbed at 56 days and has minimal tissue reaction.

POLYMER	BREAK RESISTANCE (MPa)	KNOT RESISTANCE (MPa)	ELONGATION BEFORE BREAKING (%)
Catgut	310-380	110-210	15-35
PGA	760-920	310-590	18-25
PGA-PLA	570-910	300-400	18-25
PCL	654-882	300-410	67-96
PDO	450-560	240-340	30-38
PGA-TMC	540-610	280-480	26-38

Mean characteristics of the polymers, and comparation with catgut.

b) Self-Adhesions Sutures

This new kind of sutures has recently emerged with a revolutionary new system of "barbs" that allow the latching of the suture material for tissue approximation, which eliminates a long and tedious step in wound closure, the knoting to secure and ensure the union.

For the usual closure is necessary to make knots that can cause tissue ischemia; compared to the nodes, the new system provides a more uniform tension on the soft tissue as they approach and stress is distributed along the entire wound distribution; thus, placement of the sutures with latching is less traumatic to the tissues as it eliminates the need for sutures or knots interrupted.

- Bioser®: Quill ™ Self-Retaining System (Angiotech Pharmaceuticals, Inc.)
- Covidien®: V-Loc ™

Possible advantages

- Controls and distributes the tension in the tissue (continuous suture)
- Provides the same tensile strength that the interrupted sutures. Allows wound closure including the total thickness of the skin and subcutaneous fat.
- Facilitates the closure of difficult wounds, including:
  - Wound with high strength-tension
  - Sutures bag snuff
  - Wounds with complex architecture
  - Wounds with mismatched diameters

- Reduce the complications associated with the nodes
- Eliminates the extrusion, dehiscence and irritation by the knots
- Avoid potential infections caused by stitches.
- Eliminates bottlenecks tissue (tissue or tear the cheese- wire effect)
- Take the time to suture, and reducing operating room time and associated surgical risks of prolonged anesthesia
- Possibility of better patient outcomes :

  - The stress is more evenly distributed, rather than at a single point fixing
  - Possibility of better scar due to soft tissue support .
  - Avoids the complications associated with the nodes: Absence extrusion, no pain knots.
  - No visibility or palpation of joints.

- Allows new techniques :

  - New possibilities for tissue approximation for: Laparoscopic procedures, endoscopic interventions and arthroscopic procedures .
  - Alternative for circular closure.
  - Approximation of mismatched diameters.
  - Application in difficult and tight spots.
    - Easy and quick Haemostasis : sealing tight .
    - Good vascularization / No ischemia in the incision
    - Distribute the tensile strength of suture incision
    - No knots, lower the risk of possible adhesions
    - Easy to use and quick learning curve

QUILL-SRS ™

This is a suture device with a closure system having knotless small barbs on its surface, arranged on opposite sides of a central segment without barbs directions. To facilitate insertion and passage through the tissue has a needle on each end of the suture. The suture is advanced within the tissue until the barbs engage opposite. The barbs penetrate the surrounding tissue and firmly hold the suture in place.

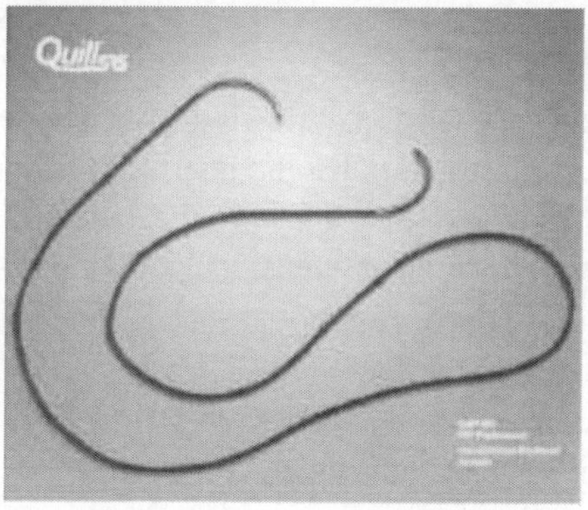

It is available in a variety of materials:

– Transparent monofilament absorbable:

Monoderm ™ (PGA-PCL):

No color (transparent).

Almost complete absorption in mass between 90-120 days. (Similar to Monocryl ™ Ethicon).

Retention / resistance to rupture in vivo:

• 62% after 7 days
• 27% after 14 days

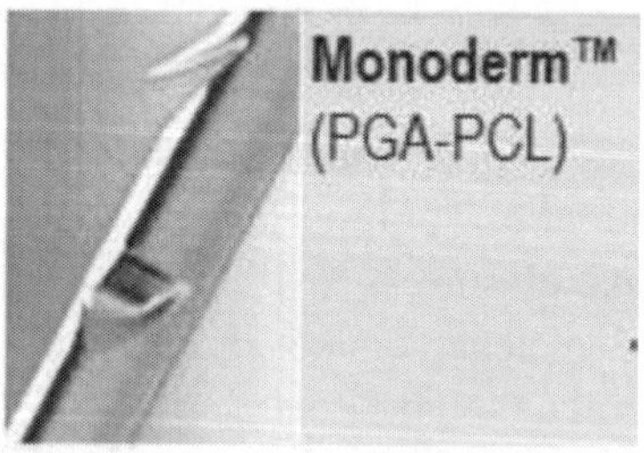

– Colored (purple) absorbable

PDO (polydioxanone)

Colored (Violet. The pigment is D & C Violet No. 2)

Absorption: Minimum to 120 days approx. Virtually complete at 180 days.

Retention / in vivo resistance to breakage by size of Quill ™ SRS:

Caliber 3-0 and above:
80% in 2 weeks
80% in 4 weeks
40% in 6 weeks

Caliber 4-0:
67% in 2 weeks
50% in 4 weeks
37% in 6 weeks

- Chemically equivalent to Ethicon PDS II ®.

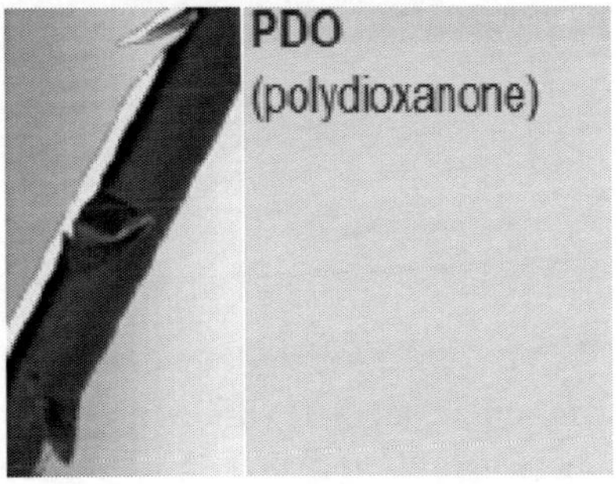

V-LOC™

The absorbable wound closure device V-Loc ™ comprises absorbable thread barbs provided with a surgical needle at one end and an end effector at the other loop. The serrated design and finished with effector loop to approximate the tissues without the need for surgical knots.

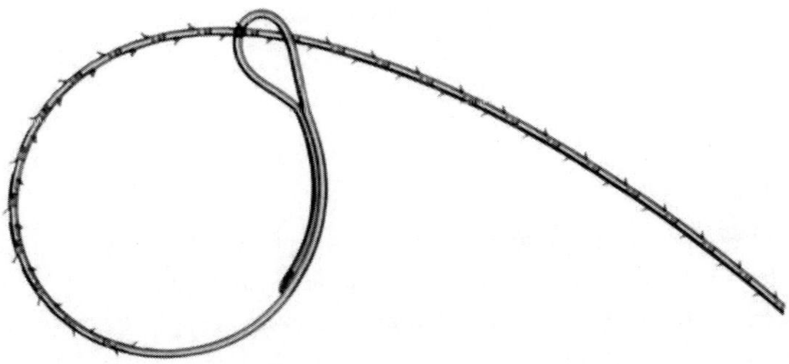

The definitions of the USP (United States Pharmacopela) to apply to the diameter absorbable wound closure device prior to the formation of the barbs. After formation of the barbs, the absorbable wound closure device shows that no toothed lower suture size. This modification reduces the tensile strength of the suture, similarly to when a knot is tied in a normal suture effect. Therefore, the tensile strength of linear traction device is comparable to the tensile strength of the USP knot for normal suturing equivalent size. Below you can see more clearly the relationship between the size of the non-serrated suture and size equivalent.

Size suturing device V-Loc ™ 180	Suture size before the formation of barbs	Normal suture equivalent size (USP) / Tensile strength (kgf)
0	1	0 / 3,90
2-0	0	2-0 / 2,68
3-0	2-0	3-0 / 1,77

It is available in two formats:

- V-LOCTM 90:

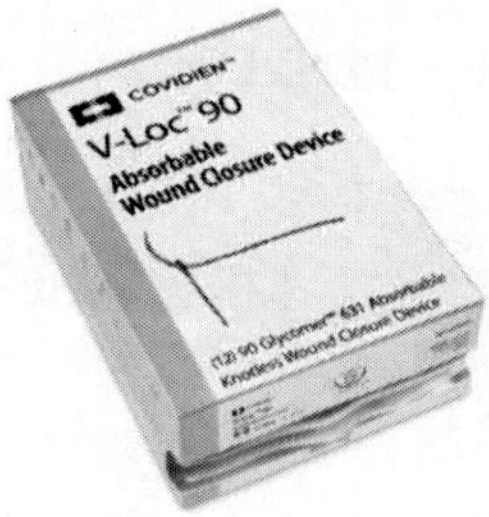

Polyglycol compound, dioxanone and trimethylene carbonate.

Colorless color and violet triangular needle cylindrical.

With an absorption profile between 90 and 110 days. With a tensile strength of 90% at 7 days and 75% at 14.

- V-180 LOCTM:

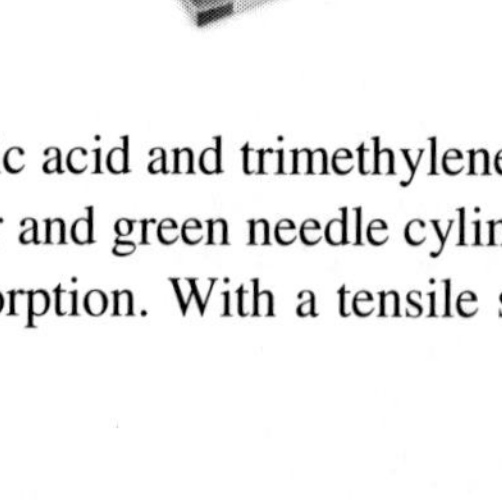

Compound copolymer of glycolic acid and trimethylene carbonate.

Colorless color for the triangular and green needle cylindrical.

With 180 days to complete absorption. With a tensile strength of 80% at 7 days, 75% at 14 and 65% at 21 days.

c) Active Sutures

Coated Sutures with Antibacterial Solutions

Surgical site infections (SSIs) are caused by many factors – some are controllable, while others are uncontrollable. Suture selection provides an opportunity to address a known risk factor for infection – bacterial colonization of the suture.

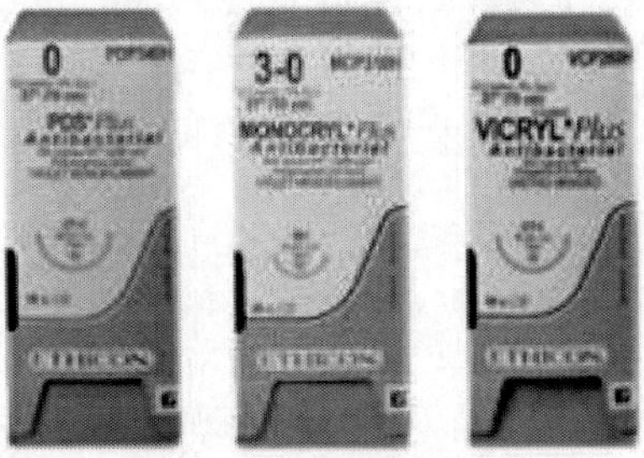

Percutaneous sutures create a conduit for bacteria to enter the wound. Additionally, like all foreign bodies, when implanted, sutures are rapidly coated with tissue protein, creating sites for bacterial colonization. This colonization can lead to biofilm formation, which can increase the difficulty of treating an infection even in the presence of antibiotics.

Coated polyglactin 910 suture, first introduced in 1974, is made of a copolymer of lactide and glucolide with a coating of polyglactic acid and calcium stearate. Because treatment of sutures by physical or chemical means can alter the handling and physical properties, the new triclosan-treated version of coated polyglactin 910 suture was tested extensively to ensure performance continuity with traditional polyglactin 910 suture. The sutures were evaluated for handling using a standard suture tie-down board and technique. This technique is utilized due to its ease of use, and has been used for many years in the quality control of sutures. Coated polyglactin 910 sutures with triclosan were compared to coated polyglactin 910 sutures (without triclosan) and were found to be indistinguishable.

One of the most efficacious biocides against Staphylococcus is triclosan (2,4,49-trichloro-29-hydroxydiphenyl ether, IRGACARE MP¨; Ciba Specialty Chemicals Corporation, Tarrytown, NY). Staphylococci are inhibited by triclosan at sub-part-per-million con centrations. Triclosan has been used successfully in treatment regimens to eliminate topical infections with methicillin-resistant bacteria. The susceptibility of the most common device-related pathogens combined with inherently low toxicity makes triclosan a favorable candidate for use in the clinical setting.

In vitro antibacterial evaluations of coated polyglactin 910 suture with triclosan using zone of inhibition assays have shown consistent efficacy results for S. aureus and S. epidermidis whether examining materials out of package, materials after 24-h or seven-day aqueous immersion, methicillin-resistant strains, knotted suture repeated inoculation materials, knotted suture deep agar inoculation materials, or materials from suture tissue pass techniques.

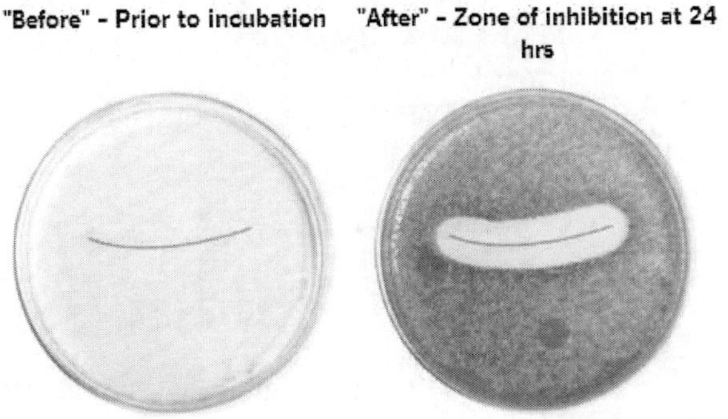

The incorporation of an antimicrobial agent into a medical device, such as a suture, can potentially affect the physical characteristics and functional properties of the device.

Plus Antibacterial Solutions are the only commercially available sutures with antibacterial protection.

- Proven in vitro to kill bacteria and inhibit bacterial colonization of the suture for 7 days or more
- Effective against pathogens commonly associated with surgical site infections, including:
  - Staphylococcus aureus
  - Staphylococcus epidermidis
  - Methicillin-resistant Staphylococcus aureus (MRSA)
  - Methicillin-resistant Staphylococcus epidermidis (MRSE)
  - Escherichia coli
  - Klebsiella pneumoniae

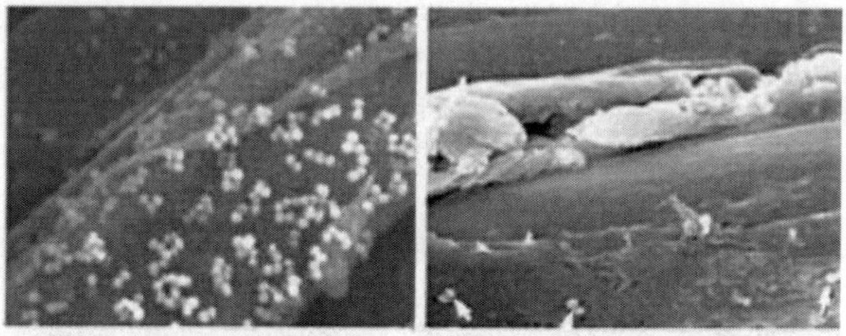

MRSA adheres to non-triclosan-treated suture (magnification 5,400X)

Only a few MRSA (arrows) adhere to triclosan-treated suture (magnification 5,260X)

Coated VICRYL™ Plus Antibacterial (Polyglactin 910) Suture

Coated VICRYL Plus Antibacterial (Polyglactin 910) Suture is a synthetic absorbable, sterile, surgical suture composed of a copolymer made from 90% glycolide and 10% L-lactide. The suture contains Irgacare MP (triclosan), a broad-spectrum antibacterial agent.

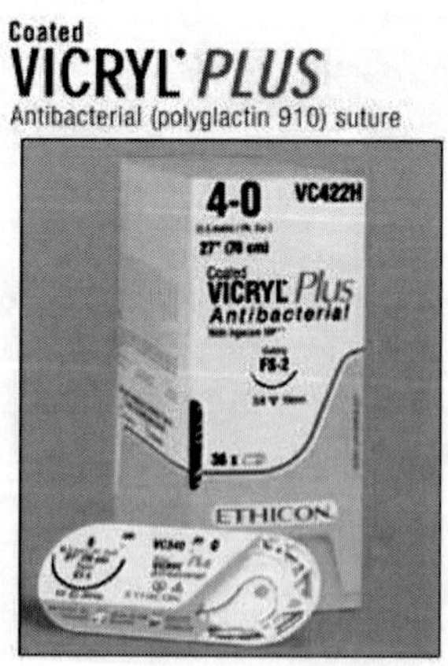

Coated VICRYL Plus Antibacterial Suture is indicated for use in general soft tissue approximation and/or ligation, except for ophthalmic, cardiovascular and neurological tissues.

MONOCRYL™ Plus Antibacterial (Poliglecaprone 25) Suture

MONOCRYL Plus Antibacterial (poliglecaprone 25) monofilament suture is a monofilament synthetic absorbable surgical suture prepared from a copolymer of glycolide and epsilon-caprolactone. The suture contains Irgacare MP (triclosan), a broad-spectrum antibacterial agent.

MONOCRYL Plus Antibacterial sutures are indicated for use in general soft tissue approximation and/or ligation, but not for use in cardiovascular or neurological tissues, microsurgery or ophthalmic surgery.

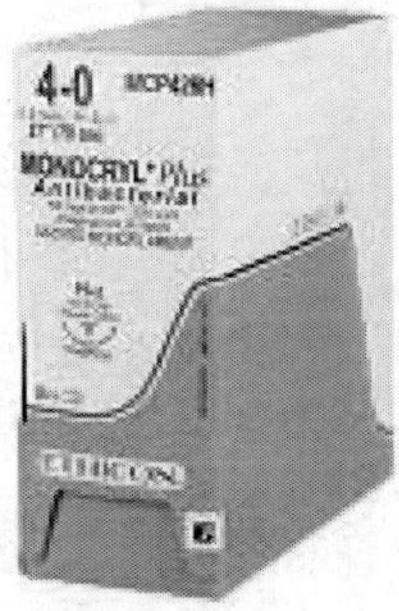

PDS™ Plus Antibacterial (Polydioxanone) Suture

PDS Plus Antibacterial (polydioxanone) monofilament synthetic absorbable suture is prepared from the polyester, poly (p-dioxanone). The suture contains Irgacare MP (triclosan), a broad-spectrum antibacterial agent.

PDS Plus Antibacterial sutures are indicated for use in soft tissue approximation, including use in pediatric cardiovascular tissue where growth is expected to occur and ophthalmic surgery (other than contact with cornea and sclera). PDS Plus Antibacterial suture is not indicated in adult cardiovascular tissue, microsurgery, and neural tissue. These sutures are particularly useful where the combination of an absorbable suture and extended wound support (up to six weeks) is desirable.

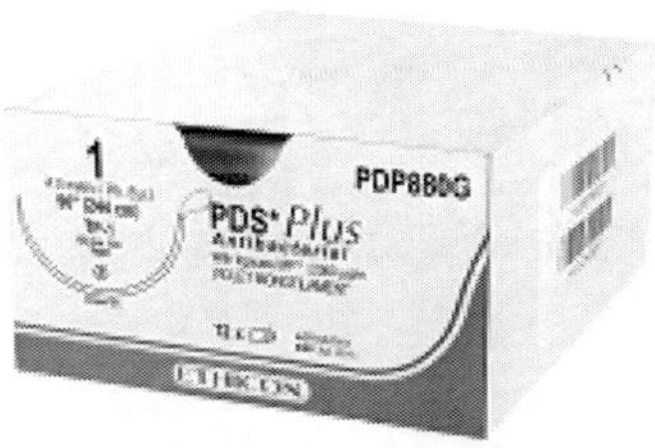

2) Use in Meshes

Biomaterials are defined as all reachable in the market as metals and alloys, ceramics, carbon derivatives, polymers and biological tissues, which have been used and are still used in surgical materials. Only some of these are the components used today in the prosthesis of hernia surgery. Types of prostheses:

- Reticular Prosthetics, useful for placement in a tissue-tissue interface.
- Laminar Prosthetics, best for placement in direct contact with the visceral peritoneum.
- Composite Prosthesis, those that can be placed on all interfaces, but its design is developed in order to be placed in a tissue interface and in turn interface visceral peritoneum.

Tipos	Absorción	Biomaterial
Reticulares	Nonabsorbable	PPL, polyester
	Partially absorbable	PPL/Polyiglactin 910 PPL/ poliglecaprone
	Absorbable	Polyglactin 910, Ac. polylactic
Laminares	Nonabsorbable	PTFE-e, silicone, polyurethane
	Absorbable	Porcine intestinal submucosa
Compuestas	Nonabsorbable	PPL/PTFEe, PPL/ polyurethane
	Absorbable	PPL with polyethylene glycol, Ac. Polylactic, Polydioxanone or Cellulose

New classification: Classification Klinge 2012

Clase	Denominación	Subtipo	Características
I	macroporous	Ia	monofilament
		Ib	multifilament
		Ic	mixed or polymer
II	microporous	IIa	monofilament
		IIb	Multifilament
		IIc	mixed or polymer
III	Special Features		pore design, anti-adhesion barrier...
IV	Meshes with film		
V	3D Meshes		
VI	Biological meshes	VIa	Non Cros-linked
		VIb	Cross-linked
		VIc	Special Features

a) Hibrid Synthetic Meshes

There are different types of meshes after the combination of these resorbable synthetic elements. They are braided prosthesis, multifilament, soft, flexible, moldable, biodegradable (resorption by hydrolysis at 3-6 months with a progressive reduction of prosthetic mass and its tensile strength).

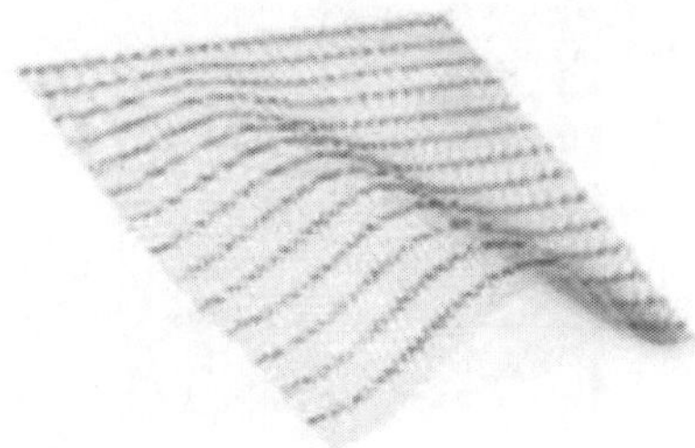

Ultra-pro mesh (Partially Absorbable Lightweight Mesh).

Composite prosthesis:

Composite or composite mesh prosthesis are arising from the fusion of two layers of material. Despite the wide range of brands available, most of these meshes are using three basic materials such as polypropylene, polyester and PTFE, used in combination with additional material, such as titanium, omega 3 fatty acids, poliglecoprone 25 hyaluronate PVDF and others.

The parietal surface or component is strengthening, made with PTFE-e, polypropylene or polyester, designed to induce fibroblastic reaction; the second component is a visceral or anti-adhesion barrier can be chemical (resorbable) with very poor reactivity composed of layers of collagen, polyethylene glycol and glycerol (Parietex Composite ®, Parietene Composite ®), hyaluronic acid and carboxymethylcellulose (Sepramesh ®), oxidized regenerated cellulose (Proceed ®) or acids omega 3 (C-QUR Edge ®, C-Cor ®), and the physical nature (nonabsorbable) such as PVDF (Dynamesh ®), silicone (Microval Intro ®, Surgimesh WN non-Adherent ®), polyurethane (PU PL-99 ®, HI-TEX ® Endo IA), titanium (Ti-Mesh ®) or PTFE-e itself (Bard Composix LP ®, Intramesh T1 ®, Dual Mesh ®) .

This type of prosthesis is especially useful in cases where repair conditions the direct contact of the mesh with the visceral peritoneum and large hernia defects or creating temporary laparostomas to prevent compartment syndrome, besides being habitual use in the laparoscopic approach.

PROCEED® Surgical Mesh.

Parietex® Composite.

b) Absorbable Synthetic Meshes

A polymer composed of the esters of polyglycolic acid (Dexon ®), or a copolymer derived from the synthesis of the latter with the lactic acid (polyglactin 910, Vicryl ®).

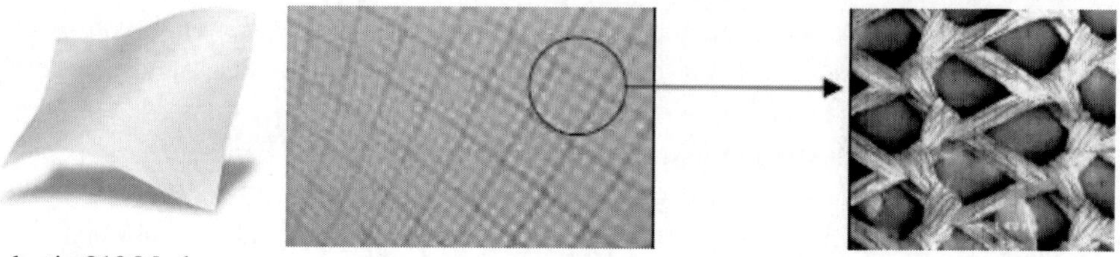

Polyglactin 910 Mesh.

We have developed a new prosthesis composed of a combination of biocompatible synthetic biopolymers polyglycolic acid (67%) and trimethylene carbonate (37%) (PGA-TMC). It has a hydrophilic microporous structure, which makes optimal for absorbing and releasing water based products at the site of implantation. Reabsorbed at 6 months, available with the design of "plug" for inguinal hernia (hernia Gore Bio-A plug ®) and flat mesh of various sizes (Gore_Bio A tissue reinforcement ®).

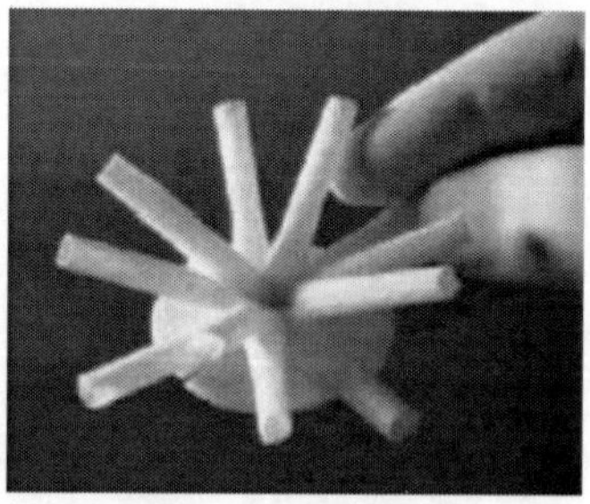

Gore Bio-A (PGA-TMC) plug.

c) Absorbable Grips in Meshes (Selfgripping- Meshes)

Application of polylactic acid in the field of abdominal wall surgery, has developed multiple advances as sutures and surgical mesh. Innovations are not only fit, also in coatings and connections with other materials.

According to this, one of the novel uses of polylactic acid is his inclusion as part as meshes for hernia repair. Specifically, PLA is the other component, joined polypropylene monofilament, of a new partially absorbable mesh, low molecular weight, which is recently utilized in abdominal wall surgery.

We are referring to a composite mesh designed with a thin PP-layer hugely less polypropylene (only 10% of the total) and an additional coating of appendices made of poly-L-lactic acid (PLA), which accounts in total for 90% of the mesh weight (Fig. 1-2) The PP-PLA mesh is bi-layered which manage reducing drastically the size of the pores and a thick layer of only PLA.

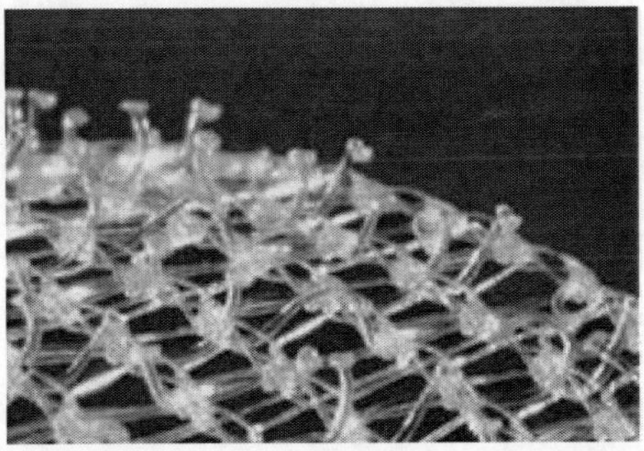

Composite bi-layered mesh made of PP monofilament and PLA.

Inner surface of the mesh with PLA grips in relief.

The appendices of polylactic acid adopting a shape as micro hooks and in this way can be inserted between the muscle fibers, like the Velcro® effect (Fig.3). The absorbable polylactic micro hooks in the PP structure provide tissue-gripping properties of the mesh during the following 12 months.

Diagram about the way of gripping ability of PLA appendices.

This composite mesh born with the aim of find a system for tension free repair in surgical treatment of hernias, more complete that the only idea of close the hole, literally, with a plug. Instead of that, this bi-layered lead obtained a great extensive coverage of the all treated area.

The mesh Parietene ® Progrip ® (Sofradim, Group Covidien, Trévoux, France) is presented in three modes, two of them elliptical shape, specially designed to fit the anatomy of the inguinal region, with a self-adhesive flap adjustable cord around (Fig 4), either left or right. The third variant is a flat plate 15x9 cm.

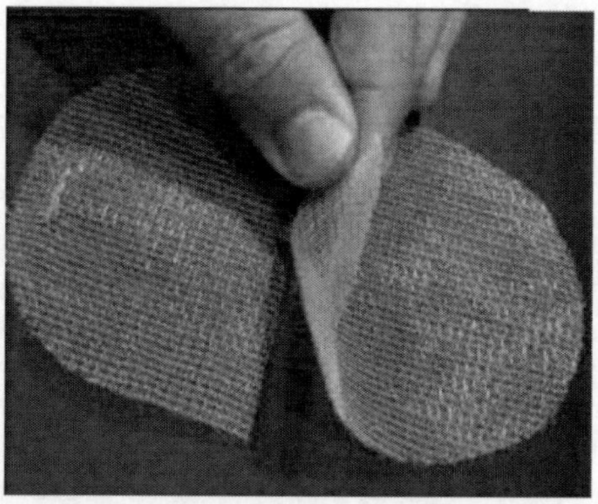

Photograph of the adherent flap of meshes anatomically designed for inguinal region.

The main advantage of this technique is the lack of additional stitches. They are not necessary for fixing the mesh because of the appendices of PLA behaves like self-fixation device allowing secure attachment from its inner surface (Fig. 5), with less risk of mobilization of the hernia sac under the prosthesis.

Aspect of the mesh once collocated after inguinal hernia repair.

That issue is in direct relation with the low incidence of chronic postoperative pain, complication often founded after surgical hernia repair that sometimes had a poor solution. Actually, since recent prospective studies gave clear evidence of the prevalence of worrying chronic pain after polypropylene mesh implantation, the design of prosthesis with an improved biocompatibility appears as a major concern.

As a matter of fact, this decrease of postoperative pain depends on the low inflammatory response which this kind of partially absorbable meshes produces thanks to a minor strange body reaction. This affirmation has been widely observed in several studies published about the use of self-gripping mesh in hernia repair, which report that after adjunction of PLA, the presence of lymphocyte was slightly and significantly reduced than the founded in histological analysis of cases where only polypropylene was used, arguing for a lowered immune response. On the other hand, the macrophage infiltration remained less important with PP-PLA meshes when compared with the both kinds of PP-meshes. This finding suggests that addition of PLA material seems to reduce inflammation in spite of its degradable feature.

These results are possible thanks to polylactic acid is an absorbable material which is completely disappeared at 12 months after surgical intervention, reducing to a half the molecular weight of the whole mesh. Data has been corroborated by electrical microscope studies (Fig.6).

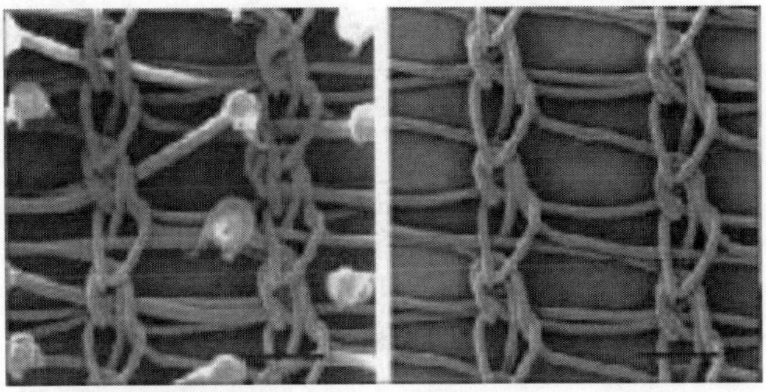

Image (EM) of mesh after PLA absorption.

The only negative point consists in a delayed adhesion of the mesh to muscle and peritoneum, as the material needed more than 1 month to reach the adhesiveness of polypropylene; but the absence of recurrence at 2 years in the longest case series published until today, is promising, and suggests that the self-gripping properties of the PLA micro hooks independently provide satisfactory mesh fixation at least equivalent to other fixation devices.

References

[1] Sajid MS, Leaver C, Baig MK, Sains P. Systematic review and meta-analysis of the use of lightweight versus heavyweight mesh in open inguinal hernia repair. *Br. J. Surg*. 2012 Jan; 99(1):29-37.

[2] Bellon JM. Proposal for a new classification of prostheses used in the repair of abdominal wall hernial defects. *Cir. Esp*. 2005 Sep; 78(3):148-51.

[3] Coda A, Lamberti R, Martorana S. Classification of prosthetics used in hernia repair based on weight and biomaterial. *Hernia*. 2012 Feb; 16(1):9-20.

[4] Bilsel Y, Abci I. The search for ideal hernia repair; mesh materials and types. *Int. J. Surg*. 2012; 10(6):317-21.

[5] Junge K, Klinge U, Prescher A, Giboni P, Niewiera M, Schumpelick V. Elasticity of the anterior abdominal wall and impact for reparation of incisional hernias using mesh implants. *Hernia*. 2001 Sep; 5(3):113-8.

[6] Lopez-Cano M, Barreiro Morandeira F. Prosthetic material in incisional hernia surgery. *Cir. Esp*. 2010 Sep; 88(3):152-7.

[7] Binnebosel M, von Trotha KT, Jansen PL, Conze J, Neumann UP, Junge K. Biocompatibility of prosthetic meshes in abdominal surgery. *Semin. Immunopathol.* 2013 May; 33(3):235-43.

[8] Vega-Ruiz V, Garcia-Urena MA, Diaz-Godoy A, Carnero FJ, Escribano Moriana A, Velasco Garcia M. Surveillance of shrinkage of polypropylene mesh used in the repair of ventral hernias. *Cir. Esp*. 2006 Jul; 80(1):38-42.

[9] García-Ureña MA, Vega Ruiz V, Díaz Godoy A, Báez Perea JM, Marín Gómez LM, Carnero Hernández FJ, et al. Differences in polypropylene shrinkage depending on mesh position in an experimental study. *Am. J. Surg*. 2007 Apr; 193(4):538-42.

[10] Poussier M, Deneve E, Blanc P, Boulay E, Bertrand M, Nedelcu M, et al. A review of available prosthetic material for abdominal wall repair. *J. Visc. Surg*. 2013 Feb; 150(1):52-9.

[11] Klinge U, Klosterhalfen B. Modified classification of surgical meshes for hernia repair based on the analyses of 1,000 explanted meshes. *Hernia*. 2012 Jun; 16(3):251-8.

[12] Amid PK. Classification of biomaterials and their related complications in abdominal wall hernia surgery. *Hernia.* 1997;1:15-21.

[13] Goulart et al., Hernioplasty in One-Day Surgery: result of 228 self-adhesive prosthesis. *Ambulatory surgery* july 2012.18:14-17.

[14] García Ureña M.A, Hidalgo M, Feliú X,. Velasco M.A, Revuelta S, Gutiérrez R, Utrera A, Porrero J. L, Marín M, Zaragoza C. Multicentric observational study of pain after the use of a selfgripping lightweight mesh. *Hernia* (2011) 15:511–515 DOI 10.1007/s10029-011-0811-3.

[15] Pedano N, Pastor C, Arredondo J, Poveda I, Ruiz J, Montón S, Molina, M, Hernández JL, Open tension-free hernioplasty using a novel lightweight self-gripping mesh: medium-term experience from two institutions. *Langenbecks Arch. Surg*. 2011 DOI 10.1007/s00423-011-0831-6.

[16] Suárez Grau JM y cols. Postoperative pain and surgical time in inguinal hernia repair with self-gripping mesh: Experience in ambulatory surgery. *Ambulatory surgery*. July 2012.18:18-20.

[17] Kosai N, Sutton PA, Evans J, Varghese J. Laparoscopic preperitoneal mesh repair using a novel self-adhesive mesh. *J. Min. Access Surg*. 2011;7:192-4.

[18] Birk JF, Hess S, García-Pardo C. Low recurrence rate and low chronic pain associated with inguinal hernia repair by laparoscopic placement of Parietex Progrip ™ mesh: Clinical outcomes of 220 hernias with mean follow up at 23 months. *Hernia* (2011)15:553-8

[19] Suárez Grau JM, De Toro Crespo M, Docobo Durántez F, Rubio Chaves C, Martín Cartes JA, Docobo Pérez F. Prevention of surgical infection using reabsorbable antibacterial suture (Vicryl Plus) versus reabsorbable conventional suture in hernioplasty. An experimental study in animals. *Cir. Esp*. 2007 Jun;81(6):324-9.

[20] Castaldi MA, Cobellis L, Fraternali F, Ardovino M, Ardovino I, Colacurci N. Biomechanical Features of Bidirectional-barbed Suture: A Randomized Laboratory Analysis. *Surg. Technol. Int*. 2014 Mar;24:45-8.

In: New Developments in Polylactic Acid Research
Editor: Courtney Winthrop
ISBN: 978-1-63463-054-2

Chapter 6

PLGA Nanosphere Platform for Nanomedical System

***Yusuke Tsukada*[1], *Hiroyuki Tsujimoto*[2], *Nobuhiko Miwa*[3], *Hiromitsu Yamamoto*[4] *and Yoshiaki Kawashima*[5*]**

[1]Pharmaceutical and Beauty Science Research Center, Hosokawa Micron Corporation, Japan
[2]Material Business Dept., Hosokawa Micron Corporation, Japan
[3]Osaka Butsuryo University and Emeritus Professor of Prefectural University of Hiroshima, Japan
[4]Aichi Gakuin University, Japan
[5]Aichi Gakuin University and Emeritus Professor of Gifu Pharmaceutical University, Japan

Abstract

We have been developing biocompatible and biodegradable polymeric micro-nanospheres with poly(D,L-lactide-co-glycolide) (PLGA), loaded with bioactive substance to create new functional medical system toward next generation. It was found that the function of such particulate system was determined by its critical micromeritic property such as particle size, shape and specific surface area and compatibility with biosubstance. The PLGA nanospheres (NS) can improve the drug absorption at the target site due to sustained stay and prolonged release of the drug, leading to the increase in pharmacological effect. It was also found that PLGA NS can be transferred into the cells by endocytosis, especially phagocytosis. This function can be applied to deliver orally nucleic acid substances such as siRNA, asORN and decoy ODN etc. by administrating drug loaded PLGA NS. Further nanomedical device such as eluting stent or balloon was created by depositing PLGA NS loaded with drug on the surface of bared stent or balloon. Although several preparation methods of PLGA NS have been proposed, we employed the Emulsion Solvent Diffusion (ESD) method developed by us for industrialization of PLGA NS process due to its simple process requiring no special

[*] Corresponding Author: Yoshiaki Kawashima, Contract Professor of Aichi Gakuin University and Emeritus Professor of Gifu Pharmaceutical University, Japan.

apparatus and easy scaling up. By using this process many kinds of functional cosmetics and hair restoration tonics containing the PLGA NS have been developed and launched on the market. Several drugs and medical devices using our PLGA NS are now under clinical trials managed by pharmaceutical companies and universities.

1. Introduction

In the recent drug developments, the discovery of a lead compound by a conventional method has become more difficult even if high-throughput screening system has been introduced. For overcoming this problem, the life cycle management by improving the existing products and the practical application of biological active substances such as nucleic acid, siRNA and decoy ODN have increasingly played an important role. In such a trend, the development of new Drug Delivery System (DDS) has become one of the crucial means for achieving the goal.

Nano-sized particles [1-7] such as liposome, micelle and polymeric nanosphere etc. aiming to drug carriers have been actively researched in recent years. Especially, we have been developing poly (D,L-lactide-co-glycolide) nanosphere (PLGA NS) loaded with bioactive substance. Many researches have revealed that PLGA NS could greatly increase the pharmacological effect of drug in the variety applications because they improved the drug adsorption at the target site due to prolonged stay and sustained release of the drug.

Several clinical trials of drugs and devices using our PLGA NS have already been conducted in Japan by pharmaceutical companies and universities. Also we have already succeeded in commercialization of PLGA NS for skincare cosmetics and hair restoration products. In this paper, we would like to introduce our several applications with PLGA NS in the nanomedical system.

2. Preparation Method and Application for Practical Use of PLGA NS to Prepare Nanomedical System

2.1. Characteristics of PLGA NS

PLGA, a copolymer of lactic and glycolic acids which are organism components, is gradually hydrolyzed to the monomers (lactic and glycolic acids) in human body. Subsequently, the monomers are decomposed to the water and carbon dioxide which finally excreted by tri-carboxylic acid cycle. Hence, PLGA has been studied for many years in wide variety of applications as a safety DDS material.

As a successful DDS product using PLGA, the subcutaneous injectable formulations of the LH-RH analogue long-term sustained release PLGA microcapsule (20~30μm) [1] used for the treatment of the prostate and breast cancer in the world is well known [8-10]. This product had changed the inconvenient daily subcutaneous injection of the conventional parenteral solution to the once-1, 3, 6 month administration, leading to the significant improvement of the patient's quality of life (QOL). In those formulations, the release rate of drug from the microcapsule can be adjusted by the PLGA's molecular weight and the molar ratio of lactic acid and glycolic acid which can control the hydrolysis rate of PLGA [8-10]. This is one of the powerful advantages for PLGA as a drug carrier material.

It must be interesting to clarify the change in characteristics of PLGA particle with decreasing the size of particle. When the particle size of PLGA particulate system was changed from micro to nano level, unexpected change in the behavior of PLGA NS was found in vivo. Firstly, the absorption and adhesion of particles to biological membranes such as the mucosal layer become stronger dramatically due to the significant increase of the specific surface area, and the permeability is also improved. For this reason, the drug absorption can be remarkably improved in combination with the effect of drug sustained-release.

Furthermore, the PLGA NS can be applied as a DDS for nucleic acids because the NS obtains the ability of intracellular drug delivery by the endocytosis [11-13]. Accordingly, it becomes possible to deliver nucleic acids to the targeted tissues and transport into the cells efficiently protecting them from the decomposition by enzymes in the human body, so many researches on those applications using PLGA NS have been reported [14, 15].

2.2. Preparation of PLGA NS by Emulsion Solvent Diffusion Method

Several methods [16-18] such as Emulsion Solvent Evaporation method, Phase Separation method, Salting-out method and Spherical Crystallization method etc. have been proposed for preparing PLGA NS. Especially, we employ Emulsion Solvent Diffusion (ESD) method [4-7] developed from Spherical Crystallization (SC) method developed by Kawashima et al. and published in Science in 1982 [19]. The SC method can produce the drug micro-nanoparticles by controlling the crystallization and agglomeration of drug crystals in the simple system. The ESD method is developed by coexistence of polymer in the SC method in order to obtain the polymeric nanosphere encapsulating drug. The fundamental principle of ESD method is the self-emulsification caused by the natural convection (Marangoni phenomenon) generated by the difference of surface tension between a water-miscible organic solvent like acetone and water (Figure 1). When such organic solvent dissolving both the PLGA and drug is mixed with water, the nano-sized pseudo-o/w droplet is formed instantly by the self-emulsification. By rapid mutual diffusion of solvent and water occurs at the same time, PLGA and drug in the droplets are precipitated, leading to obtain the solid PLGA NSs encapsulating drug. Hence, the ESD method which never requires strong mechanical energies used in other common process such as the ultrasonication, high-pressure emulsification and others for producing various nanoparticles, is advantageous to encapsulate the nucleic acid and peptide easily decomposed by such powerful energies.

Furthermore, in the ESD method, the surface of PLGA NS can be modified with the biocompatible cationic polymer chitosan which can improve the NS's retention and adhesion to biological membranes, by using the aqueous phase dissolving chitosan. The chitosan-modification can improve the encapsulating efficiency of drug into NS and the cellular uptake of NS [11, 12, 20].

The almost spherical PLGA NSs with the narrow particle size distribution can be prepared by the ESD method. The particle size control is also possible. Although the emulsion droplets momentarily solidified, they are more likely to coalesce comparing to other droplets whose surfaces were stabilized by surfactants. It has confirmed experimentally that it's important for the particle size control in ESD method to limit the frequency of coalescence [21]. For instance, the decrease of viscosity in the aqueous phase by increasing its temperature or the increase in stirring rate could improve the droplet dispersion, prevent droplet coalescence and resulted in a smaller apparent particle size of PLGA NS. In addition, the average particle diameter of PLGA NS was directly proportional to the cube root of PLGA concentration in the solvent. This was because PLGA molecule precipitated in the emulsion droplet had a certain constant density. By adjusting the operating parameters of ESD method depending on the rule experiences, PLGA NS can be prepared to pass through a membrane filter with 0.2μm pore size used for sterilization as shown in Figure 2. In this way, we have already established GMP manufacturing technology and platform for producing PLGA NS with the level of sterility assurance according to pharmaceutical guidelines, and actually the sterile PLGA NS encapsulating drug used for the clinical trials have been produced at our platform.

2.3. Design of Nanocomposite Particulate System by Using Powder Technologies

Nanopaticles are generally difficult to handle and tend to aggregate in manufacturing process and during storage. If nanoparticles aggregate, they cannot demonstrate their original abilities as nanoparticles in the human body. Especially, in the case of PLGA having low grass-transition temperature which varies depending on the molecular weight of PLGA and molar ratio of lactic acid to glycolic acid, the aggregation and coalescence of PLGA NS is likely to occur because of the low thermal stability. For overcoming those problems, several powder technologies using mechanical particle composite method [22-25] can be applied because PLGA NS is a rigid material compared to other DDS carrier particles like liposome, micelle, emulsion and etc. existing as a droplet. The powder technologies enable us to prepare the micro-sized PLGA nanocomposite particle which PLGA NSs are dispersed in a matrix composed of water-soluble excipient such as mannitol (Figure 3). Namely, preparing the nanocomposite particle can not only improve the storage stability of NSs but also allows handling NSs in the same way for preparing solid dosage forms such as tablet, capsule, dry powder inhalation (DPI) and so on.

3. DDS Formulation of PLGA NS Prepared by ESD Method and Resultant Pharmacological Performance

3.1. Oral Nanocomposite DDS with Hormone such as Insulin and Calcitonin

The development of the transmucosal administratable DDS for peptide hormone etc. with short biological half-life alternative to frequent injection has been desired. This chapter describes the development of oral peptide DDS using the PLGA NS.

Oral administration of zinc-insulin encapsulated PLGA NS to the rats loaded with sugar by subcutaneous administration of glucose has been found to be useful [26].

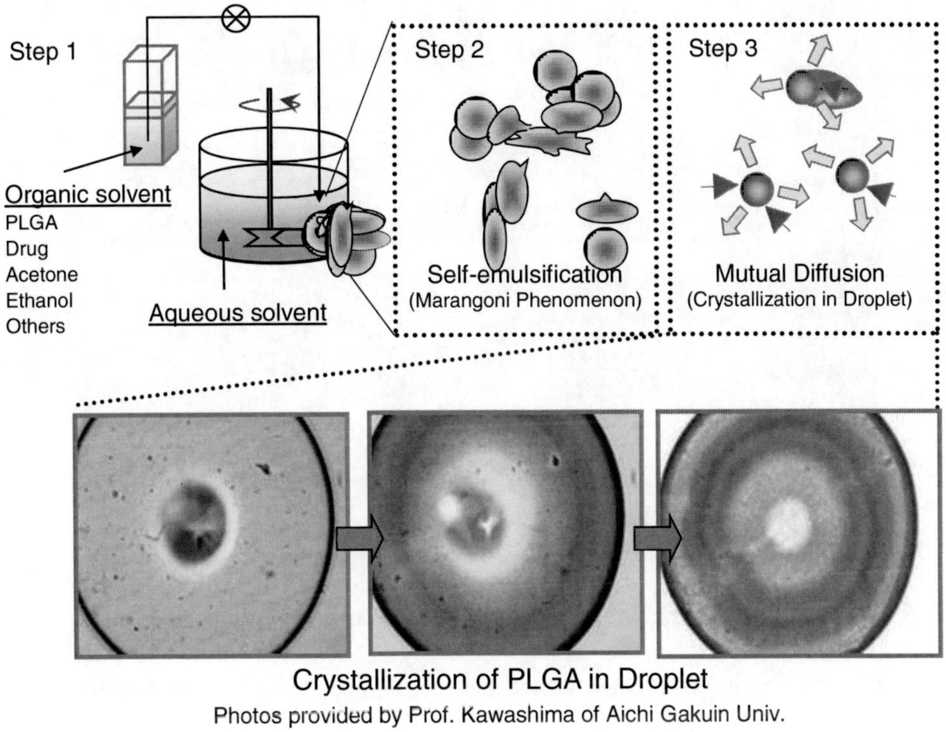

Figure 1. Preparation of PLGA nanosphere by Emulsion Solvent Diffusion Method.

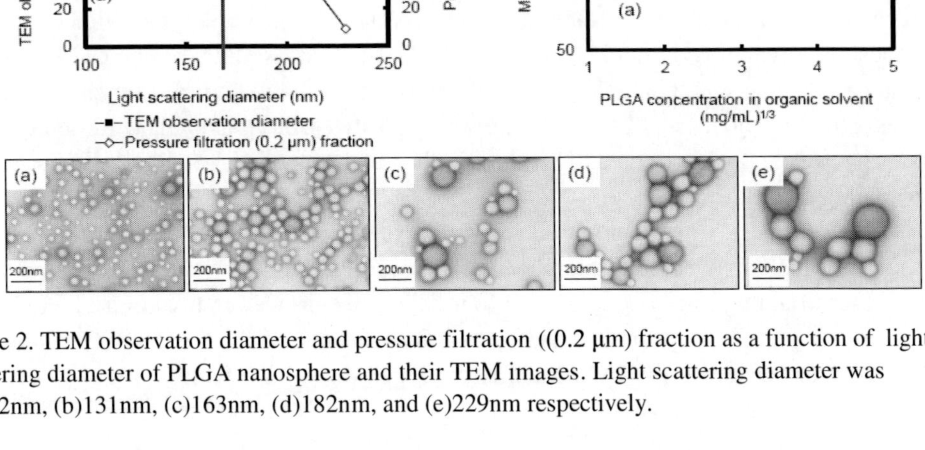

Figure 2. TEM observation diameter and pressure filtration ((0.2 μm) fraction as a function of light scattering diameter of PLGA nanosphere and their TEM images. Light scattering diameter was (a)102nm, (b)131nm, (c)163nm, (d)182nm, and (e)229nm respectively.

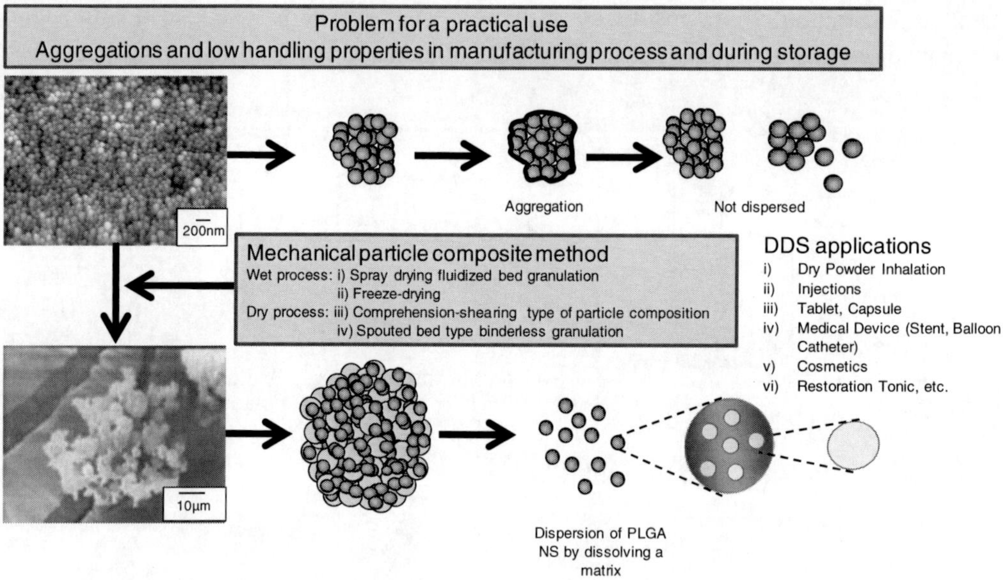

Figure 3. PLGA nanocomposite particle design by mechanical particle composite method for a practical use.

The PLGA NS was prepared by the phase transition method; the mixture of the microcrystalline suspension of zinc-insulin and methylene chloride solution dissolving PLGA was emulsified by the ultrasonication, after that the obtained solution was dispersed into the petroleum ether.

While almost no reduction of the blood glucose level appeared in the intraperitoneal administration with zinc-insulin solution, the PLGA NS maintained the lower level over several hours. In the report, the result suggested that the PLGA NS released insulin effectively because of PLGA NS's adhesion to the gastrointestinal mucosa and the increase of the insulin's absorption by the cellular uptake of NS.

On the other hand, we established the oral peptide delivery system by developing the mucoadhesive PLGA NS by using the ESD method [4, 27, 28]. This concept is to improve the oral adsorption of peptide by enhancing the carrier's adhesion to and penetration into the mucosal layer and the retention at the adsorption site.

Prepared PLGA NS was calcitonin encapsulated PLGA NS modified with the mucoadhesive polymer such as chitosan, polyacrylic acid and sodium alginate. The surface modification could be manipulated by changing zeta potential and particle diameter of NS. In particular, chitosan modified PLGA NS showed a positive zeta potential, while non-coated PLGA NS was negatively charged. The particle diameter tended to be slightly larger due to the CS coating layer.

The evaluation of the penetration behavior of PLGA NS containing fluorescence substance into the removed intestinal tract of rat was carried out by tracer method using confocal laser scanning microscope. It was found that the surface modification of PLGA NS with chitosan significantly promoted the penetration and mucoadhesive of PLGA NS. Oral administration of calcitonin-PLGANS modified with chitosan (500 IU/kg) to rat significantly reduced the blood calcium level than that of calcitonin solution, and the reducing effect was maintained for 36 hours (Figure 4) [27].

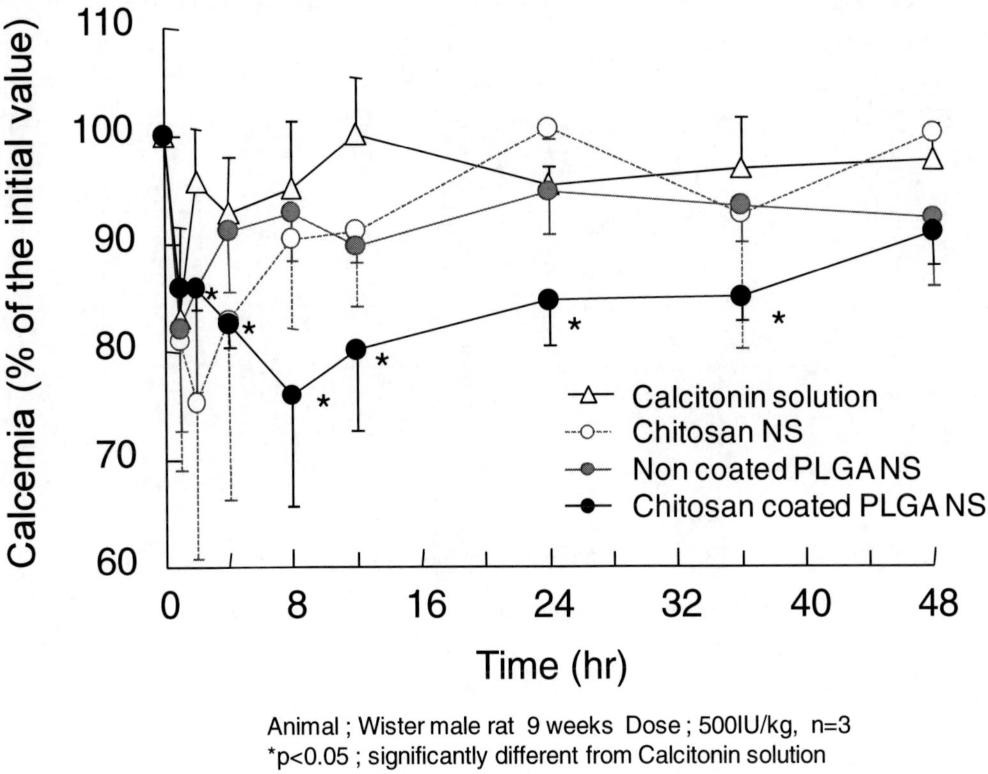

Figure 4. Pharmacological effect of calcitonin loaded PLGA NS by oral administration to rat.

The reducing effect of blood calcium level and its prolongation in the case of administrating unmodified PLGA NS was not so significant compared to that of the chitosan modified PLGA NS.

3.2. Pulmonary Nanocomposite DDS with Hormone or Antiasthmatic Drug

Pulmonary administration has attracted much attention as an alternative to intravenous or subcutaneous administration for drug delivery due to its advantageous properties such as large adsorption area of alveoli comparable to instetinal tract, thin epithelial membrane for the ease of absorption, lower enzymatic degradation activity.

We have established the insulin-DPI system using inhalation device such as Jethaler® alternative to the invasive and conventional subcutaneous injection [22-25]. In the development, several kinds of composite particle involving of insulin loaded PLGA NS have been prepared (Figure 5).

One is the soft matrix composite paticle prepared by using a spray drying fluidized bed granulator (Agglomaster®) with sugar alcohol such as mannitol (c.a. 4μm in diameter) and insulin-PLGA NS [23]. The second one is the ordered mixture of coarse carrier lactose (c.a. 4μm in diameter) adsorbed with the PLGA NS processed by using the comprehension-shearing type of particle composing machine (Mechanofusion®) [22]. The third one is soft matrix composite particle prepared by the spouted bed type binderless granulation method using freeze-dried PLGA NS and fine lactose powder (c.a. 2μm in diameter) grinded by a jet mill [25].

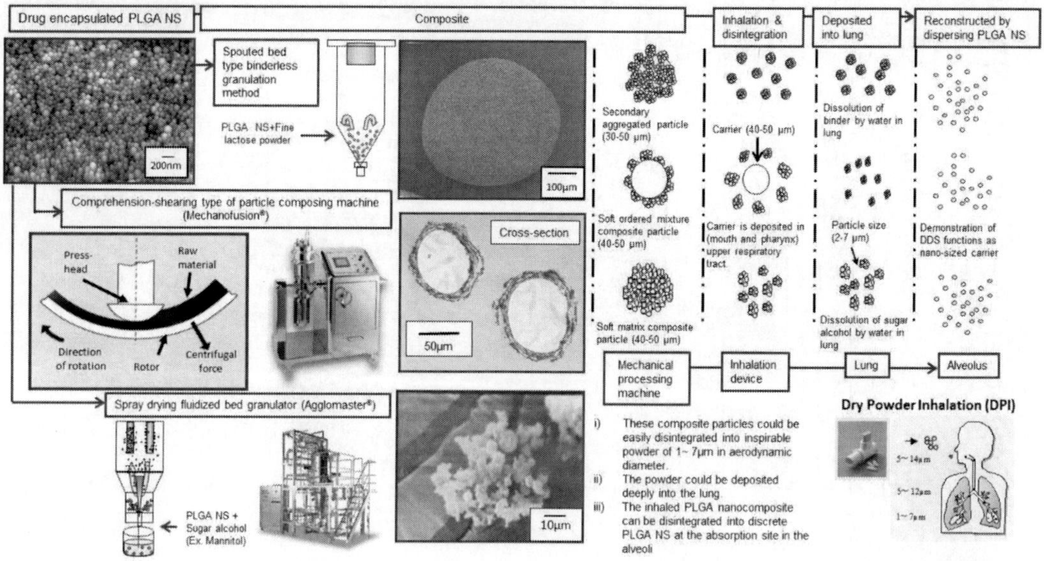

Figure 5. Dry powder inhalation system using PLGA nanocomposite particles.

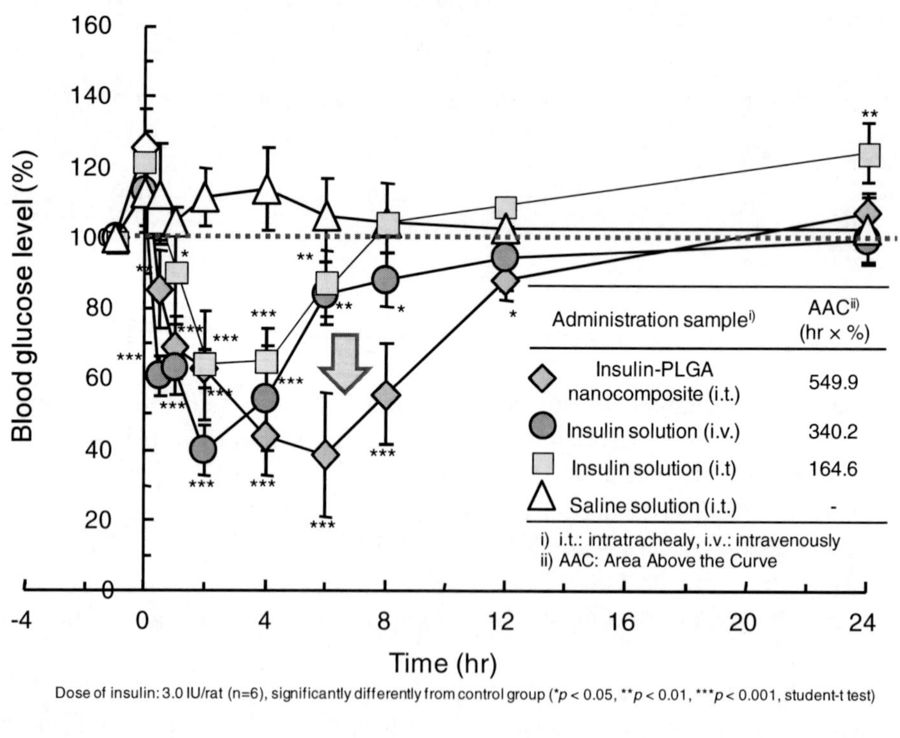

Figure 6. Pharmacological effect of insulin-PLGA nanocomposite particle administered intratrachealy to rat.

Those composite particles were designed to be easily disintegrated into inspirable powder of 1~ 7μm in aerodynamic diameter, and such dispersed powder could be deposited deeply into the lung. In vitro inhalation property of the insulin-PLGA nanocomposite was evaluated with a cascade impactor showing higher respirable fraction of over 40% than that of 10% of the freeze-dried PLGA NS alone. Then original PLGA NS suspension could be readily reconstructed by dispersing the PLGA nanocomposite into distilled water. This finding suggested that the inhaled PLGA nanocomposite can be disintegrated into discrete PLGA NS at the absorption site in the alveoli, leading to the instantaneous and sustainable systemic action of insulin by the DDS performance of PLGA NS. Actually, area above the curve (AAC) of blood glucose level of PLGA nanocomposite intratracheally administration was approximately 3.4 and 1.6 times higher than that of intratracheally and intravenously administration of insulin solution in normal rat, respectively (Figure 6) [22]. Additionally, in the evaluation using beagle dogs intratracheally administered the PLGA nanocomposite under spontaneously breathing, AAC of blood glucose level was approximately 3.5 times higher than that of subcutaneously administration of insulin solution [24].

Pharmacokinetics of inhaled PLGA NS has been revealed by pathological and histological examinations[29, 30]. After FITC (green fluorescent marker) conjugated PLGA NS were gently administered into the trachea of rat, PLGA NS were uptaken by type 1 alveolar epithelium cells immediately and remained at least 24 hours after administration. In addition, they were also gradually taken up by alveolus macrophages 30 minutes after administration and remained at least 24 hours as well.

Furthermore, PLGA NS was gradually found in the blood at 30 minutes after administration. Electron microscope observations indicated that there was a threshold in particle size for PLGA NS to be absorbed in the body, which was 100nm or smaller. These findings suggested the followings: The larger NS with several hundred nm can be suitable for topical application such as targeting delivery of antiasthmatic drug into trachea macrophages, and the smaller NS under 100nm for systemic applications such as sustained blood levels of insulin as described above.

3.3. Oral Nanocomposite DDS with Nucleic Acid (NF-κB Decoy Oligonucleotides) for Colon Delivery

Nucleic acid has been widely used as a potential tool for innovative therapy. Especially, one of the most promising agents as the next-generation medicine is nuclear factor-κB (NF-κB) decoy developed by Prof. Morishita [31, 32]. The NF-κB decoy is a synthetic double-stranded oligonucleotides which can block the activation of genes mediating inflammatory bioactive cytokine such as interleukin-1 (IL-1) and tumor necrosis factor α (TNF-α).

We have already developed the chitosan modified PLGA NS containing NF-κB decoy. Applying the phenomenon that forms a polyion-complex composed of anionic NF-κB decoy and cationic chitosan by their electrostatic interaction, NF-κB decoy can be loaded on the surface of NS modified with chitosan as well as inside of NS (Figure 7). This particulate design enables to overcome the technical hurdle for a practical use such as drug release control and increase of drug payload in product [20, 33].

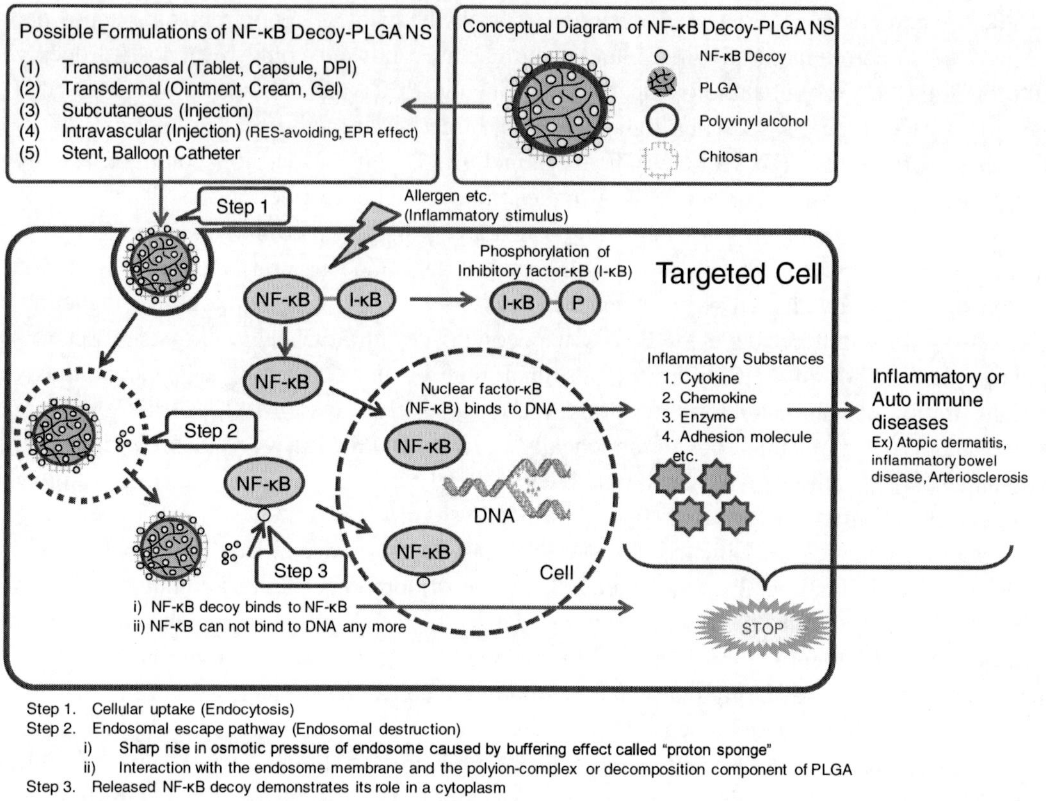

Figure 7. NF-κB decoy delivery to the targeted cell by PLGA nanosphere.

The mechanism of NF-κB decoy delivery to the cell with PLGA NS is explained as follows.

Firstly, the positively charged NF-κB decoy-PLGA NS has high affinity and adsorption to the cell surface negatively charged due to the phosphate groups. Then, the adsorbed PLGA NS is promptly uptaken into the cells by endocytosis. After that, NF-κB decoy loaded on the PLGA NS surface is released preferably, thereafter encapsulated NF-κB decoy is released gradually as the decomposition of PLGA.

Both the PLGA NS and NF-κB decoy can escape from the endosome by the following hypotheses: i) Sharp rise in osmotic pressure of endosome caused by buffering effect called "proton sponge" of the amino group in chitosan molecule, ii) Destruction of the endosomal membrane due to the interaction with the membrane and the polyion-complex or decomposition component of PLGA. In this way, the released NF-κB decoy shows original role as described above in the cell. This NF-κB decoy-delivery system using PLGA NS have been ever applied for several developments such as injection and oral formulations for inflammatory bowel disease, transdermal formulations for atopic dermatitis and the medical devices (stent and balloon catheter) used for treating stenosis of blood vessel.

In particular, the usefulness of the NF-κB decoy-PLGA NS by oral administration for the inflammatory bowel disease has been reported [15]. Although the naked NF-κB decoy was decomposed within 2 hours in an acidic medium (pH1.2) or pH 7.4 buffer medium containing DNase I corresponding to a gastric condition and an intestinal condition respectively, our developed PLGA NS system were able to prevent those decomposition significantly.

Daily oral administration of NF-κB decoy-PLGA NS in the ulcerative colitis (UC) model rat significantly improved dextran sulfate sodium-induced diarrhea, bloody feces, shortening of colon length, and myeloperoxidase activity. Furthermore, NF-κB decoy-PLGA NS was specifically deposited and adsorbed on the inflamed mucosal tissue of the model rat. These results suggested that PLGA NS modified with chitosan provide an effective means of colon-specific oral NF-κB decoy delivery in UC. This achievement led us to the next stage for the development of oral dosage forms in order to verify the applicability to the human clinical trial.

In the development, the formulation technologies for the enteric tablet and seamless capsule containing the NF-κB decoy-PLGA NS had been established [34]. The developed tablet or capsule passed the "Disintegration test of an enteric-coated tablet and capsule" defined in the Japanese Pharmacopoeia.

The results of in vitro release tests in acidic (pH1.2) and weak alkaline (pH7.4) buffer solutions revealed that those preparations had suitable characteristics as oral enteric-formulations for targeting colon's inflammation area. Actually, the preparations demonstrated the effectiveness for the chronic UC model mouse. Additionally, in the manufacturing process, the tablets and capsules could be manufactured within the acceptable degradations of PLGA and NF-κB decoy. Also we verified that the process was capable of scale-up for mass-production under GMP regulation.

3.4 Transdermal Nanocomposite DDS

Many kinds of nano-sized particles have been researched as carrier system on the transdermal drug delivery [35-37]. In this field, we also found that PLGA NS could enhance the skin permeability of drug. Figure 8 showed the photographs taken over several hours after applying an aqueous dispersion of PLGA NS encapsulated with coumarin-6 as a fluorescent marker on the human skin compared to that of applying coumarin-6 solution alone. The coumarin-6 could permeate the skin much better and sustained longer in the tissue by encapsulation in PLGA NS although the concentration of coumarin-6 in the PLGA NS suspension was 1/100,000th of that of coumarin-6 solution [38]. Following the test result, we evaluated the enhancement effect of skin permeation of fat-soluble vitamin C derivative (VC-IP, Ascorbyl Tetraisopalmitate) quantitatively using human skin. VC-IP is an active whitening ingredient widely used for cosmetic products. Figure 9 indicated the changes over time in the amount of reduced Vitamin C in the dermis after applying an aqueous dispersion of PLGA NS encapsulated with VC-IP compared to that of applying a VC-IP emulsion on the skin samples. Vitamin C could be detected 2 hours after applying the PLGA NS, and the cumulative amount of Vitamin C after 4 hours could attain 10 times of that of applying VC-IP emulsion. Further, the Vitamin C was detected until 48 hours after applying the PLGA NS, whereas no longer detected within 7 hours after applying the VC-IP emulsion. These results showed the benefits of using PLGA NS to supply an active ingredient immediately and continuously to the deep skin [38].

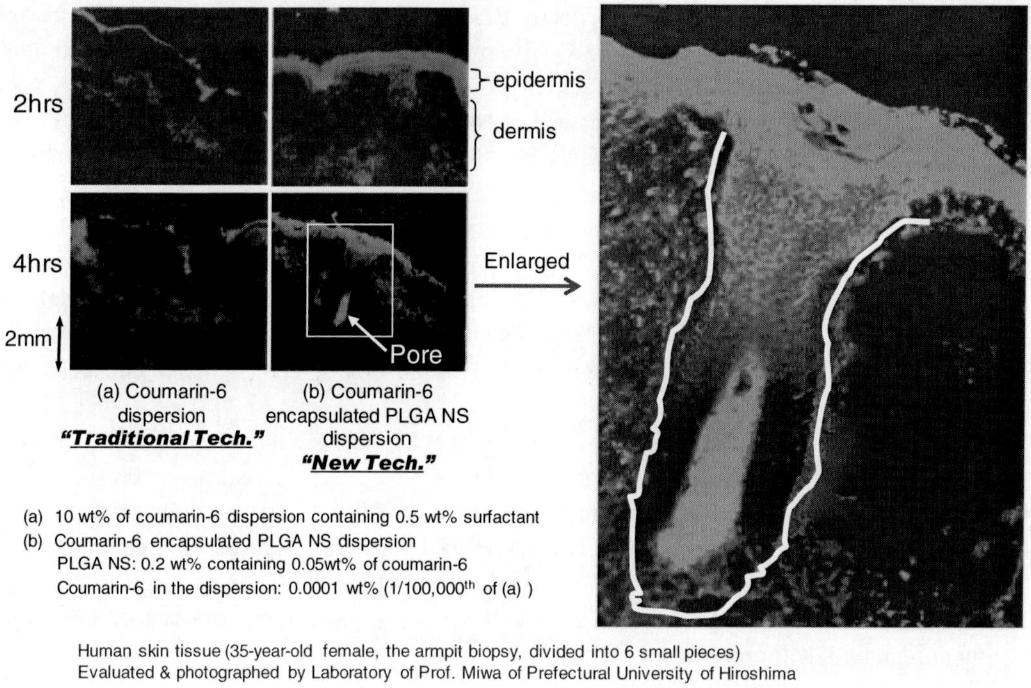

Figure 8. Enhancement of skin permeability of drug by encapsulation with PLGA nanosphere.

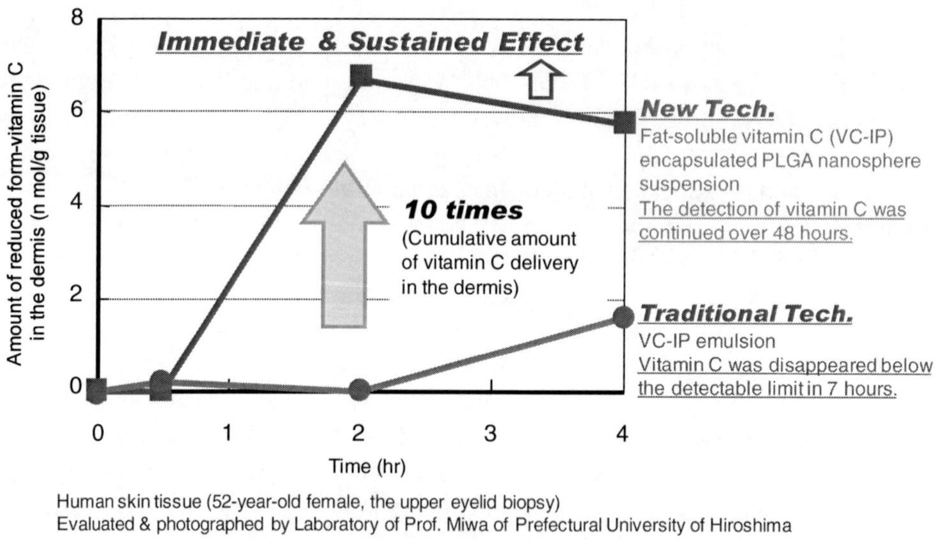

Figure 9. Ascorbic acid (reduced form-vitamin C) converted from VC-IP in the dermis.

The enhancement effect of skin permeability of drug by PLGA NS was examined in more detail using three-dimensional cultured human skin model (LSE-high, TOYOBO). The experimental result using fluorescein (FL) as a model drug was shown in Figure 10. The amount of permeated FL of 24 hours after applying the aqueous suspension of PLGA NS encapsulated with FL showed more than 2.5 times higher than that of applying FL solution alone. Interestingly, the amount of permeated FL was increased 3.4 times by mixing blank-PLGA NS (encapsulated with no FL) in FL solution [39].

This phenomenon, however, was not observed when the non-bioabsorbable polystyrene nanoparticles having similar size and zeta potential were dispersed in FL solution [39]. These results showed that PLGA NS could enhance the skin permeability of drug depending on not only the size effect but also its material characteristics. One of the reasonable reasons for such enhancing effect is α-hydroxy acid like lactic and glycolic acid generated by the decomposition of PLGA, which has peeling effect of the stratum corneum as a skin barrier function. It has been suggested that they also weakens the adhesion between keratinocytes. Therefore it is estimated that the chemical peeling effect plays an important role to enhance the skin permeability of drug.

In addition, the partition coefficient of FL to skin model from dispersion increases since the ratio of molecular / ionic form of a weakly-acidic FL increases with the decrease of the dispersion's pH from 7.4 to 6.5 by the lactic and glycolic acid, leading to improvement of skin permeability of FL.

Similar result has been obtained in the evaluation using tranexamic acid (TA) used as an active whitening ingredient in cosmetic field. The amount of permeated TA through the skin model in the case of mixing blank-PLGA NS in TA solution showed 2.5 times higher than that of TA solution alone.

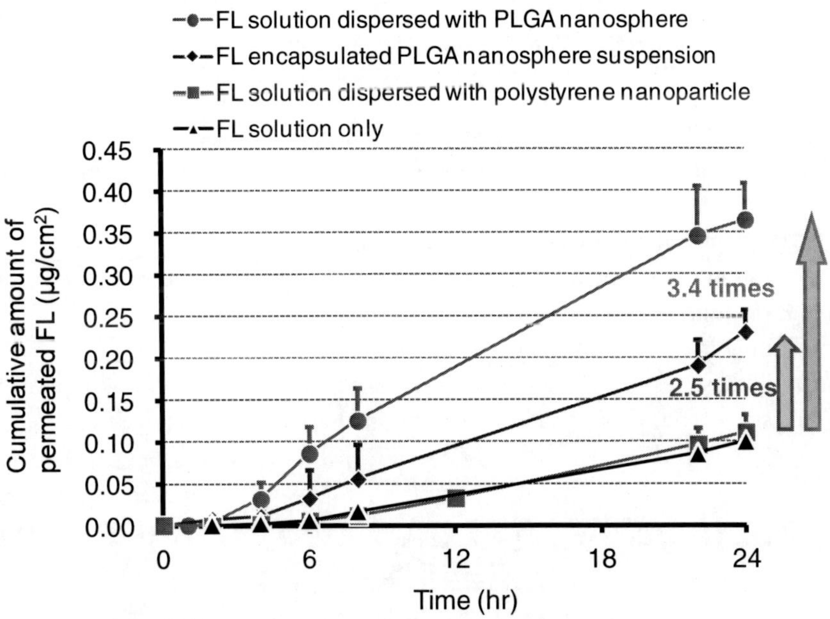

Three-dimensional cultured human skin model (LSE-high, TOYOBO)
PLGA nanosphere (179 nm, -25 mV) and
polystyrene nanoparticle (198 nm, -46 mV): 1.0 wt%
FL encapsulated PLGA nanosphere: 0.2 wt%, FL concentration: 2.5×10^{-3} wt%
Franz-type diffusion cell: 0.64 cm^2, Administered donor fluid volume: 0.5mL,
Receptor: PBS (pH7.4), 32°C

Figure 10. Effect of nano-sized carrier particle on the permeation of fluorescein (FL) through skin model.

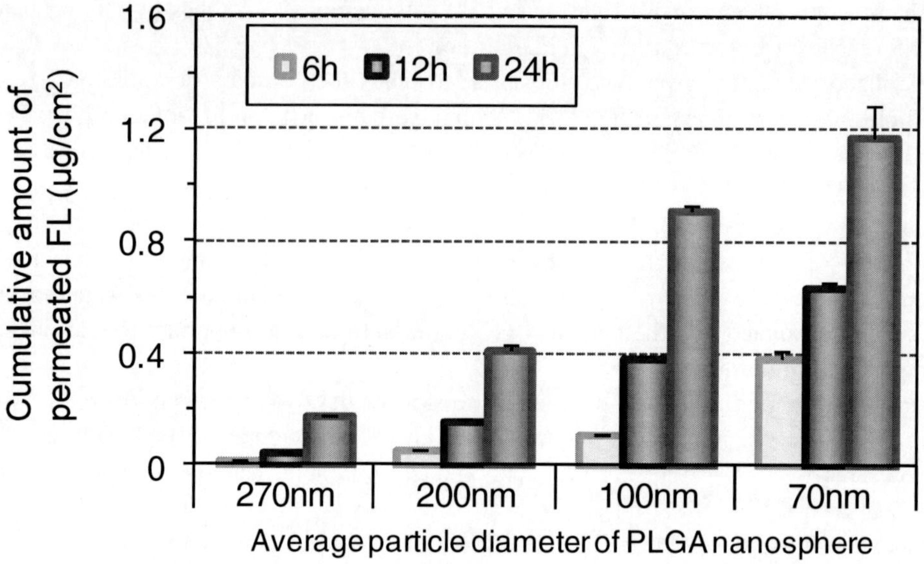

Three-dimensional cultured human skin model (LSE-high, TOYOBO)
FL encapsulated PLGA nanosphere: 0.2 wt%, FL concentration: 2.5×10^{-3} wt%
Franz-type diffusion cell: 0.64 cm^2, Administered donor fluid volume: 0.5mL,
Receptor: PBS (pH7.4), 32°C

Figure 11. Effect of particle size of PLGA nanosphere on the skin permeability of drug.

The correlation between the enhancing effect of skin permeability of drug and the particle diameter of PLGA NS has been revealed. The skin permeability of FL increased with the reduction of the particle diameter from 270nm to 200, 100 and 70nm as shown in Figure 11 [39]. Irrespective of the particle size, the pH of the dispersions was almost same value (from 7.4 to 6.8). Thus the following was considered. With the downsizing of NS, the permeability of NS to the stratum corneum was enhanced. This led to the increase of FL concentration in the stratum corneum. For this reason, the diffusion rate of FL in the skin model was increased, resulting in the improvement of skin permeability of FL.

As shown in these results, PLGA NS could considerably improve the skin permeability of drug due to its size-effect, peeling effect and sustained-release of drug.

4. Application of PLGA NS for Drug-Eluting Medical Devices

4.1. Development of Drug-Eluting Stent using a Novel Coating System

Percutaneous transluminal coronary angioplasty (PTCA) is widely known as an acceptable treatment for coronary artery stenosis. However, the usefulness of PTCA using a balloon catheter developed as a pioneer therapy has been limited by acute coronary occlusion and restenosis, which occurs in several % of patients within 24 hours and 30-40 % of patients within 3-6 month after treatment respectively.

Table 1. Past stream of development of drug-eluting stent (DES)

Generation	Feature	Product Name	Maker	Drug	Coating polymer	Platform
1st	Nonspecific antiproliferative effects on the endothelial cells	Cypher®	J&J (USA)	Sirolimus (immunosuppressive drug)	Polyethylene vinyl acetate + polybutylmethacrylate	SUS 316L
		TAXUS®	Boston Scientific (USA)	Paclitaxel (anti-cancer drug)	Polystyrene isobutylene styrene	SUS 316L
2nd	Biocompatible polymer coating	Endeavor®	Medtronic (USA)	Zotarolimus (immunosuppressive drug)	Phosphorylcholine	Co-Cr alloy
		Xience V®	Abbot Vascular (USA)	Everolimus (immunosuppressive drug)	Fluoro-polymer	Co-Cr alloy
3rd	Biocompatible polymer coating	Nobori®	Terumo (Japan)	Biolimus A9 (immunosuppressive drug) Coated onto only vascular side surface	Poly-lactic acid	SUS316L
	Complete bioabsorbable	IGAKI-TAMAI®	Kyoto Medical Planning (Japan)	None	None	Poly-L-lactic acid

Note 1: The following 3rd generation DESs have been developed; i) porous surface loaded with drug without any polymer, ii) bioabsorbable Mg alloy platform.
Note 2: Except for PLGA nanosphere, no other nano-particulate DDS-based DES has been ever developed.

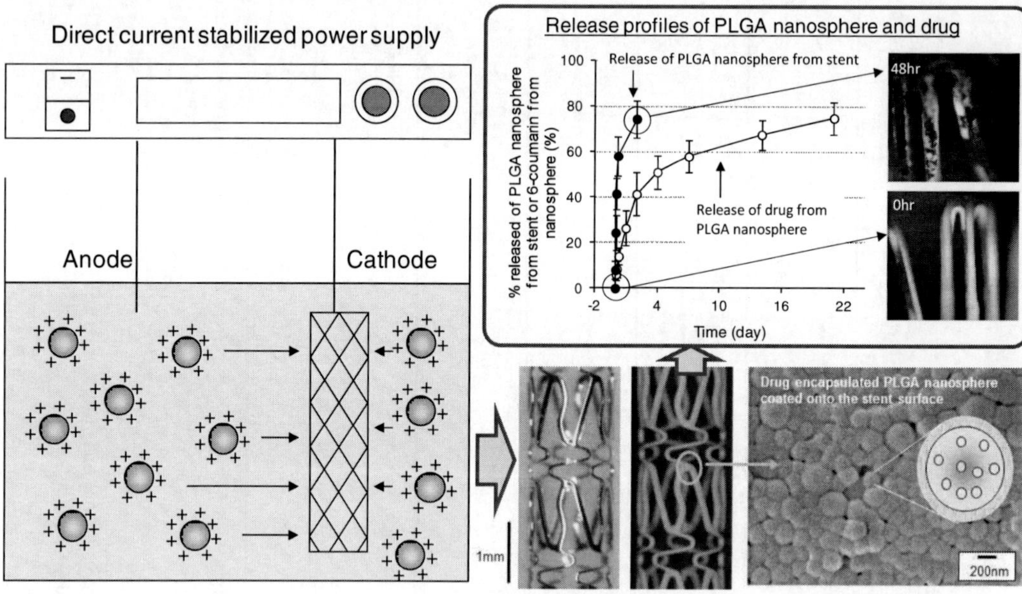

Figure 12. Electrodeposition method for coating PLGA nanosphere onto the stent and the release profiles of the prepared DES.

Then, a bare metal stent (BMS) of cylindrical mesh tube has been developed in order to prevent the above problems by supporting from inside the blood vessel. Although BMS inhibited acute coronary occlusion significantly, the rate of restenosis was still 20-30%.

The major reason of restenosis has been currently understood as an excessive growth of vascular smooth muscle cells induced by balloon injury. Therefore, the drug-eluting stent (DES) has been developed by coating BMS surface with a non-biocompatible polymer containing anti-cancer drug or immunosuppressive agent such as paclitaxel or sirolimus having the inhibition effect of cell growth respectively. Now the DES is widely used as a powerful device for reducing the restenosis rate to below 10% in PTCA.

However, recent studies [40-42] have revealed that the risk of late in-stent thrombosis (resulting in acute myocardial infarction and death) after more than 6 months of the implantation of DES became higher than that of BMS, due to the nonspecific antiproliferative effect of drug on endothelial cells and the inflammatory response derived from non-biocompatible polymer remaining permanently.

Table 1 showed the developments of next-generation DES for solving these problems.

To overcome those problems in Table 1, we have been focused on PLGA NS coated DES whose concept is to deliver efficiently drug encapsulated in PLGA NS to the targeted cells.

Desirable properties for encapsulated drugs are to suppress the excessive proliferation and migration of smooth muscle cells selectively and regenerate vascular endothelium moderately as well. Current DES polymer-based coating technologies use dip- and/or spray-coating methodology using organic solvent. These technologies are useful for coating the hydrophobic drugs such as paclitaxel and sirolimus onto the stent but not for water-soluble drugs or nucleic acid agents suited for the above described strategy regarding PLGA NS-DES. Furthermore, even if water is used as coating solvent in those technologies, it is impossible to coat PLGA NS onto the stent because of the low affinity of PLGA NS to the metal. For overcoming the problem, the electrodeposition method has been developed [43-45]

(Figure 12). In this method, cationic or anionic charged PLGA NS in water is moved and adhered to the opposite charged stent by the electrophoresis principle. This method can control the amount of coated PLGA NS by adjusting the applied voltage, energizing time and surface charge of PLGA NS. Also it can coat the PLGA NS without its deformation or decomposition.

The coating layer of PLGA NS has practically sufficient strength to avoid peeling off when the PLGA NS coated stent is mounted around the stent with being shrunk, delivered to the targeted site in the blood vessel and expanded at the site.

Additionally, the developed PLGA NS-DES can pass the Insoluble Particulate Matter Test for Injections defined in the pharmacopoeia normally used for the quality testing of the existing DES.

Then, in vitro dissolution test using saline solution of 37 °C confirmed that the coated PLGA NS was released from stent during several days and after that the released PLGA NS showed its original sustained-release property of drug (Figure 12). This release profile is thought to be appropriate for preventing the restenosis because the growth and migration of smooth muscle cells generally occur 7-14 days after stenting. As the proof of this concept, this PLGA NS-DES demonstrated by in vivo test using pigs that PLGA NS entered into the targeted cells promptly and the drug was released in the cells slowly, and the vascular endothelium was reproduced well with the same level of inflammatory reaction compared to BMS[43]. From those results, the developed PLGA NS-DES can be expected to reduce the risk of late in-stent thrombosis of current DES.

4.2. Usefulness as DDS Carriers for Application for Balloon Catheter

As with the stent, a balloon catheter applied with PLGA NS has also possibility to become an innovative medical device. Actually, the new balloon catheter coated with NF-κB decoy-PLGA NS for preventing the restenosis of arteriovenous-shunt vessel surgically formed in the upper arm of hemodialysis patients has been developed [46, 47].

As previously described, inflammation of the blood vessel is strongly involved in the vascular restenosis. Therefore, NF-κB decoy is thought to be an ideal drug suppressing the restenosis because it can adjust the expression of genes involved in this inflammation to the normal level.

Moreover, since the enhanced activity of the transcription factor NF-κB has been identified in atherosclerotic tissues of humans, NF-κB decoy is also expected to control the growth and migration of vascular smooth muscle cells.

Currently, the usefulness of NF-κB decoy to the prevention of restenosis has been found in the in vivo tests using disease model of rats, rabbits, pigs and dogs [48-53]. Additionally, the protective effect of endothelial cells in the experiments with cultured cells and the improvement of endothelium-dependent vasodilating action in the vein graft have been reported [54].

For a practical use, however, NF-κB decoy has weak points: i) it is hardly coated onto the balloon surface, ii) it is easily decomposed in human body, and iii) it does not show enough cell transfection efficiency. Then, NF-κB decoy-PLGA NS system described above has been applied for solving these problems.

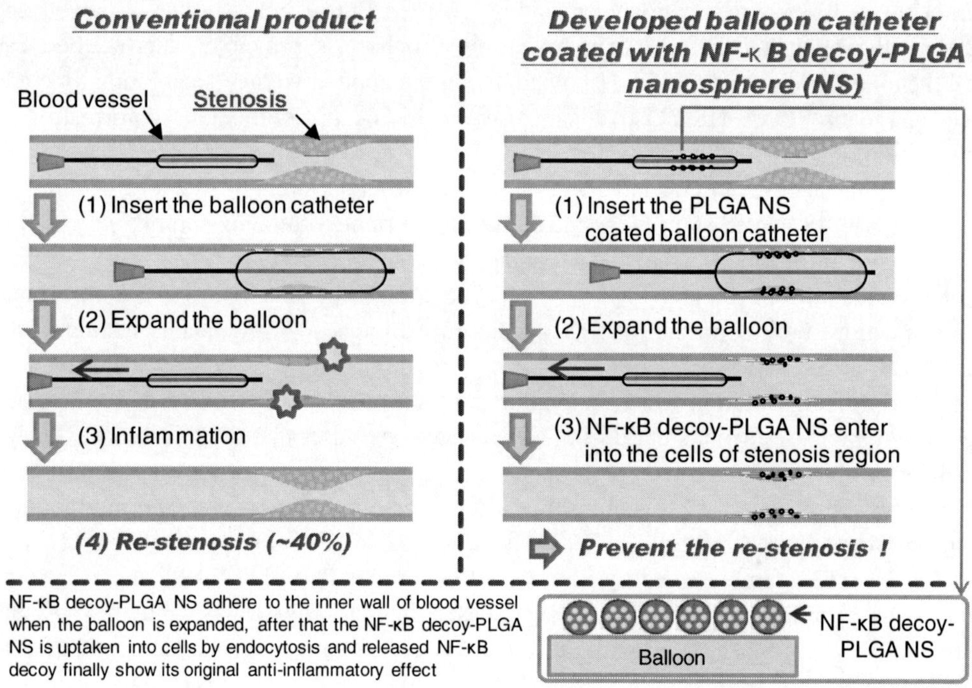

Figure 13. Conceptual diagram of the balloon catheter coated with NF-κB decoy-PLGA nanosphere for preventing the re-stenosis as a problem of the conventional product.

Using developed balloon catheter coated with NF-κB decoy-PLGA NS, NF-κB decoy-PLGA NS adhere to the inner wall of blood vessel when the balloon is expanded, after that the NF-κB decoy-PLGA NS is uptaken into cells by endocytosis and released NF-κB decoy shows its anti-inflammatory effect (Figure 13). In vivo evaluation using rabbit model has revealed satisfactory NF-κB decoy delivery into the cells and statistically significant effect for preventing the restenosis [54]. AnGes MG Inc. has truly conducted the clinical trial of this device in Japan since 2012 [47]. This device has been expected to become the world's first product using nucleic-acid substance for treating stenosis although several products coated with anti-cancer drug etc. has been already accepted for coronary artery stenosis.

5. Application of PLGA NS for Cosmetics

5.1. Development of Whitening and Anti-aging Cosmetics

Related to the enhancing effect of skin permeability of VC-IP by PLGA NS as described in 3.4, the usefulness of applying VC-IP loaded PLGA NS on inhibiting the DNA damage of skin biopsies by UV radiation has been evaluated[38]. The VC-IP loaded PLGA NS emulsion was applied to human skin biopsies, and an accelerated UV comparable to the extent of normal daily exposure was then irradiated.

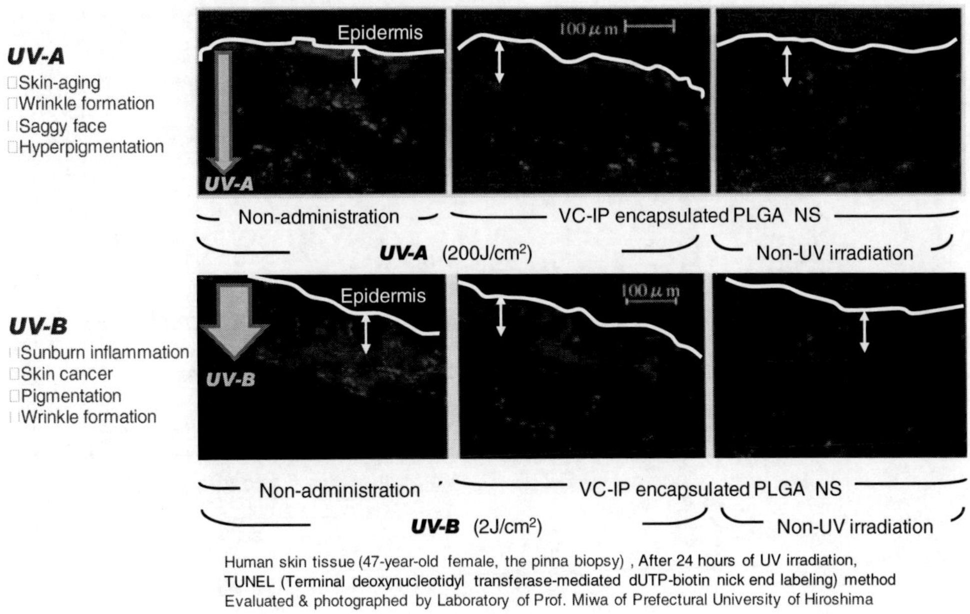

Figure 14. Protective effect of VC-IP encapsulated PLGA nanosphere on DNA damage caused by UV radiation.

In Figure 14, we showed the damaged DNA strands illustrated as the fluoresced areas selectively stained by the TUNEL (Terminal deoxynucleotidyl transferase-mediated dUTP-biotin nick end labeling).

The damaged area of skin samples administered VC-IP loaded PLGA NS emulsion were similar conditions to the un-irradiated skin samples. This result demonstrated that the VC-IP loaded PLGA NS emulsion effectively inhibited the DNA damage due to the anti-oxidation effect of the reduced vitamin C efficiently delivered into the skin depth by the PLGA NS.

Based on these findings, we have launched the world's first commercialization of the skincare cosmetic product containing PLGA NS named "NanoCrysphere®" in 2004 [55]. The three kinds of PLGA NS encapsulated with VC-IP, vitamin E and vitamin A derivatives respectively were blended in the product to provide the whitening and anti-aging effects. This product revealed the inhibitory effect of melanin production in the test skin by the evaluation using melanin-producing human cells (HMV-II) as shown in Figure15 [38].

Anti-wrinkle effect of this product has been also proved by using human skin biopsies [55]. In this evaluation, an accelerated wrinkle formation on the human skin by UV-A irradiation was reduced by administration of this product before the irradiation.

Furthermore, continuous use of this product by adult females demonstrated that this product could improve the beauty skin index such as freckles and wrinkles on their face and their satisfaction as well.

After this product, many DDS-functionalized cosmetics for sensitive or atopic skincare, makeup foundation, anti-acne and hair growth tonic etc. as described below using PLGA NS encapsulated with active ingredients have ever been commercialized (Figure16).

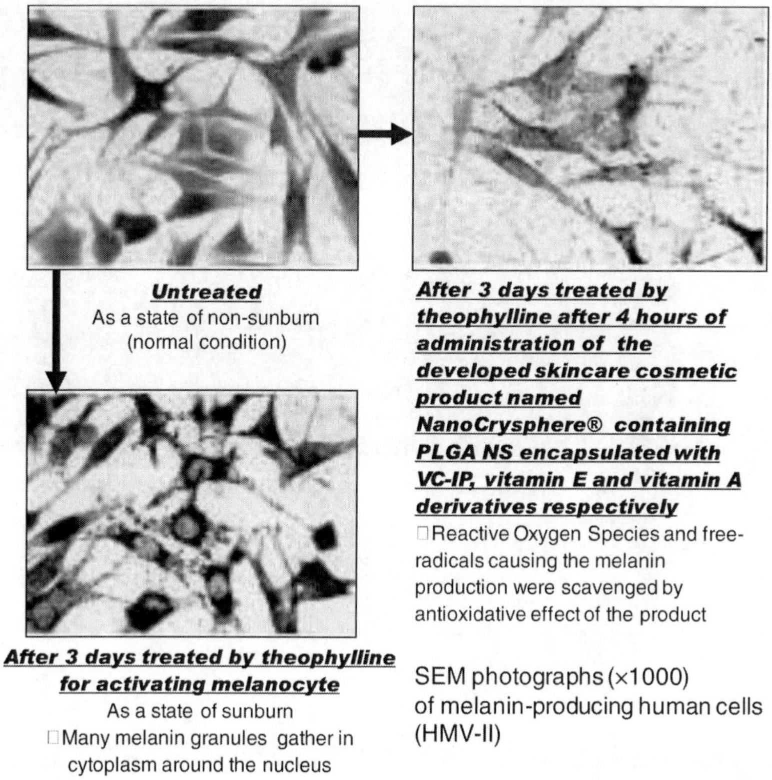

Figure 15. Whitening effect of developed skincare cosmetic product containing PLGA nanosphere evaluated by in vitro test.

(Left) Skincare cosmetic series "NanoCrysphere®"
(Left) Two dosage type essence filled PLGA NS powder and milky lotion separately
(Center) Cream, (Right) Loose powder
(Center) Skincare cosmetic series "PLGAns®"
(Upper) Pressed powder
(Lower) All-in-one gel that can be used instead of lotion, essence and cream
(Right) Hair growth tonic named "NanoImpact+®"

Figure 16. Functional skincare cosmetics and hair growth tonic containing PLGA nanosphere.

5.2. Medical Skincare PLGA NS for Sensitive or Atopic Dermatitis Skin

Stratum corneum serves to prevent the invasion of allergen or virus from outside and the water evaporation from inside to keep the skin moisture as well. It is so-called "bricks-and-mortar model [56]", a layer structure of protein-enriched corneocytes embedded in a lipid-enriched intercellular matrix. Especially, ceramide, which is the main components of the intercellular lipids, primarily plays a responsible role for skin protections as mentioned above.

The amount of ceramide in the skin of atopic dermatitis is generally known to be significantly less than that of the normal skin.

It was assumed that PLGA NS improved the autoimmune function by continuously supplying sufficient amount of ceramide or its precursors acting on the ceramide production pathway, as well as cell activating or anti-inflammatory agents to the skin by PLGA NS. In particular, the essence containing three kinds of PLGA NS loaded with sphingomyelin as a ceramide precursor, stevia extract as an anti-inflammatory agent and VC-IP as an immunoreactivity and cell activation agent respectively have been developed. In our investigation, the effectiveness of this product for atopic dermatitis patients with mild or moderate symptoms has been observed as shown in Figure 17. It was found that 29 patients (86%) were improved (remarkably: 17, slightly: 12), 5 patients were unchanged and no patient was exacerbated [39].

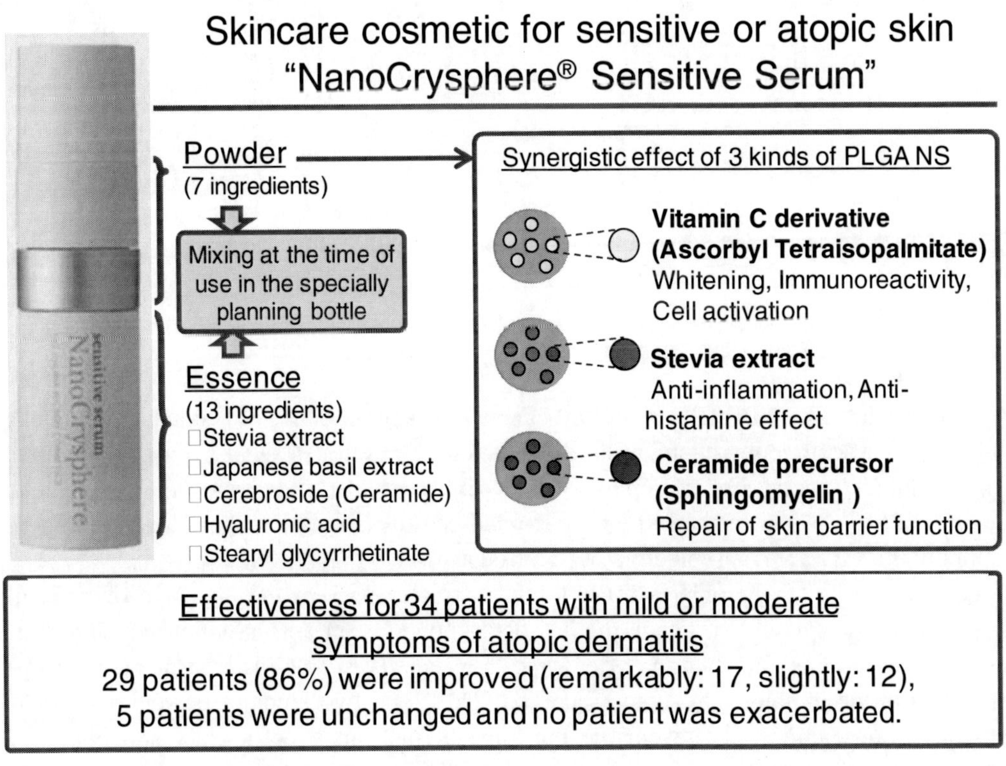

Figure 17. Application of PLGA nanosphere for treating sensitive or atopic dermatitis skin.

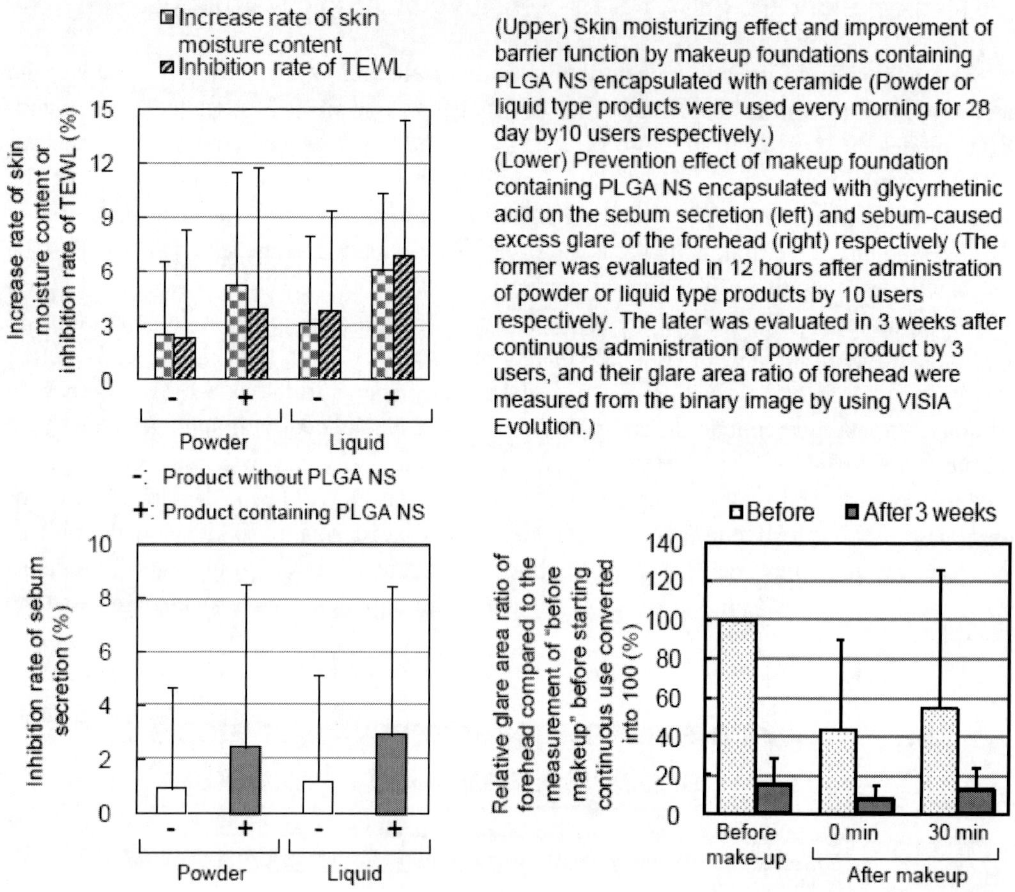

Figure 18. The usefulness of PLGA nanosphere for creating DDS-based functional makeup foundation.

5.3. Moisturizing and Breakage-Free Type Makeup Products

As represented by a BB (Beauty Balm) cream, a makeup product with skin care functions has been recently attracting attention. In this relation, the use of PLGA NS enables to create such a unique cosmetic. For examples, our developed powder and liquid type foundations containing PLGA NS encapsulated with ceramide could increase the skin moisture content and inhibit the TEWL (Transepidermal water loss) as well[*)] (Figure 18) [39].

Besides, one of the problems for a foundation is to prevent the breakage of makeup caused by sweat and sebum. Glycyrrhetinic acid known as an anti-inflammatory and anti-allergic agent inhibits an activation of enzyme (5-α-reductase) as a male hormone activator, this results in the inhibition of sebum secretion. Therefore, two kinds of powder and liquid type foundations aiming to prevent the breakage of makeup by efficiently supplying the glycyrrhetinic acid to the sebaceous gland as a targeted site in the pore using PLGA NS have been developed. Continuous use of these products for 3 weeks inhibited the sebum-caused excess glare of the forehead as an index for the breakage of makeup as shown in Figure 18 [39]. In addition to such a glycyrrhetinic acid delivery technology based on PLGA NS, high adsorption of the PLGA NS with both the skin and the foundation component is estimated to be a key factor for the success of development of the breakage-free type foundations.

5.4. Effective Anti-acnes Agent Delivery to Pores by PLGA NS

Acne is a common skin disease that many young people of 10-20 years of age mainly experience. Factors promoting the development of acne are: activation of the male hormone, increased sebum production, cornification of pores, excess growth of the *propionibacterium acnes (p-acnes)* and inflammation. Although many skincare cosmetics and medical drugs containing anti-bacterial, anti-inflammatory and keratolytic agents have been already marketed for the prevention or treatment of acne, there are still many cases of unsatisfactory therapeutic effect and adverse effect such as irritation and rubefaction as well.

For such circumstances, we created the compacted powder product with efficient anti-p-acnes agent delivery to pores by PLGA NS.

In the 20 users test, nobody improved by the usage of the compacted powder containing normally recommended concentration (0.2 wt%) of certain anti-p-acnes agent only. In addition to this, a few users suffered from irritations. On the other hand, the product containing 0.1 wt% of PLGA NS loaded with 1 wt% of the agent, which the concentration of the agent in the product was 0.001 wt%, dramatically improved the symptoms of all users without any adverse effects.

As shown in Figure 19 obtained by a certain user having dozen acnes, these acnes began to disappear after 6 days, and then almost no acne was observed after 3 weeks [57].

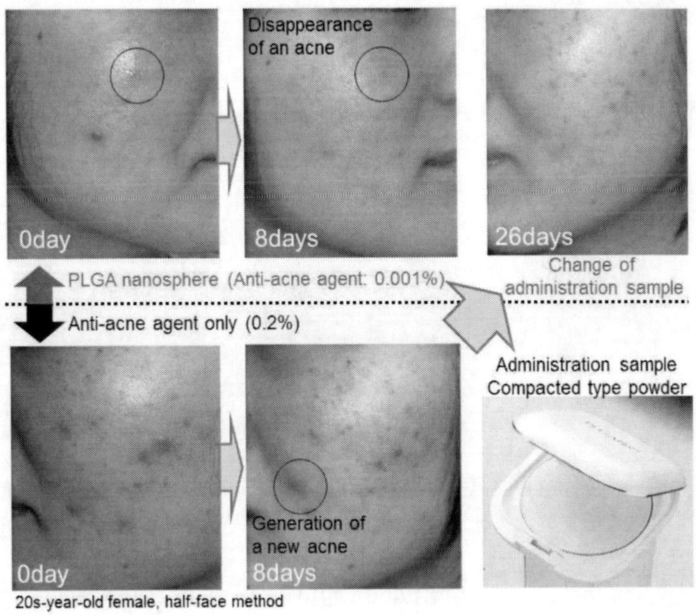

Figure 19. Improvement of face acnes by anti-acne agent delivery using PLGA nanosphere to skin pores.

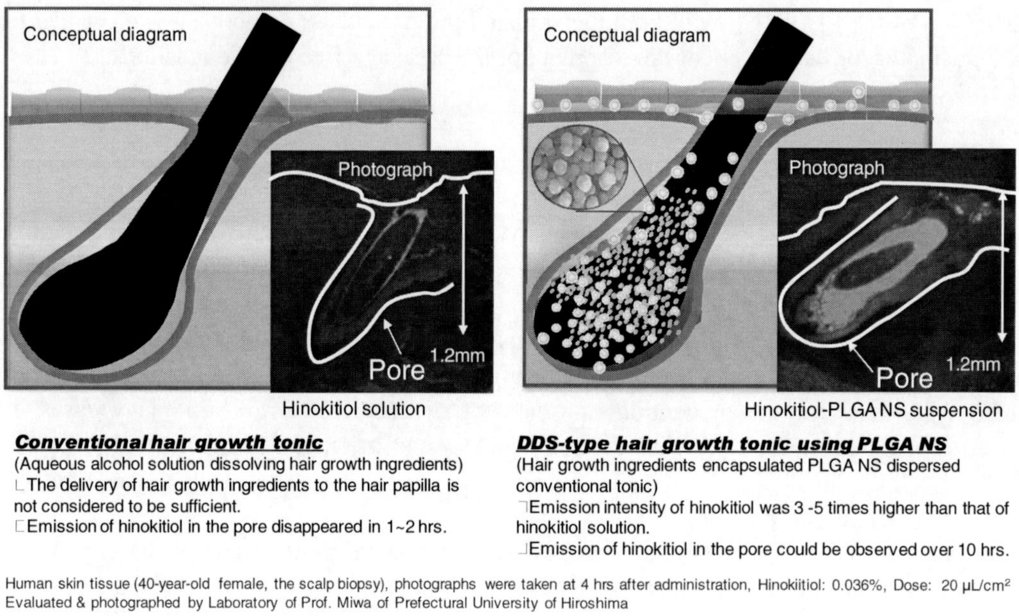

Figure 20. Development of PLGA nanosphere based-DDS type hair growth tonic.

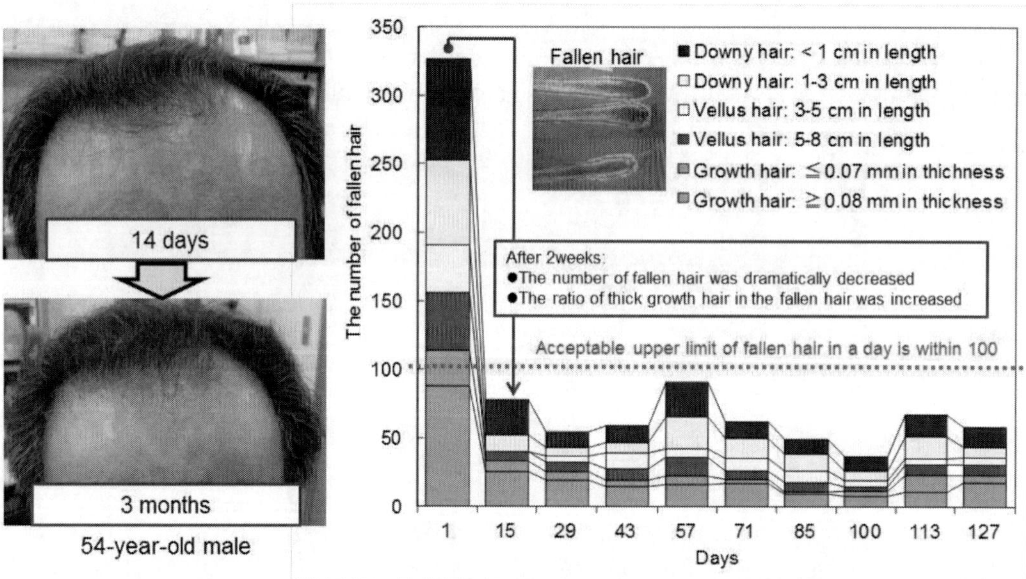

Figure 21. Hair growth effect of developed hair growth tonic containing PLGA nanosphere.

5.5. DDS-functionalized Hair Growth Tonic using PLGA NS

Many hair growth ingredients required to be penetrated to the dermal papilla cell in the depth of hair follicle for demonstrations of their own effects. Accordingly, the performance of hair growth tonic depends on the delivery ability of the ingredients to the dermal papilla cell. Current products in the market, however, which are aqueous alcohol solutions only dissolving hair growth ingredients, are not considered to deliver the ingredients efficiently. Actually, the administration of hinokitiol solution based on conventional technology to the human scalp tissue showed the emission intensity of hinokitiol in the hair follicle of 3-5 times lower than that of DDS type of hinokitiol-PLGA NS suspension proposed by us as shown in Figure 20 [58].

This is probably caused by; PLGA NS could readily pass through the hair follicle easily clogged with sebum since the particle size of PLGA NS was smaller than $1/1{,}000^{th}$ of the hair follicle diameter (approx. 0.2 mm), the liquid bridge formed in the hair follicle was restricted by the lowering of interfacial tension through the absorption of administered PLGA NS on the interface of the liquid phase (72 -> 50 dyn/cm).

Based on these findings, we have developed the hair growth tonic which PLGA NS encapsulated with blood flow acceleration, cell activation and androgen suppression agents were dispersed in a traditional type lotion.

The hair growth effect of this product has been investigated by administered to shaved back of C3H mice. Our developed product demonstrated higher hair growth effect than some marketed products. This effect has been also verified by over 50 users test [59]. Figure 21 shows the typical improvement case after the continuous use for 5 months. The number of fallen hair after 2 weeks was dramatically decreased from more than 300 to less than 100 which was acceptable upper limit of fallen hair in a day, and the ratio of thick growth hair in the fallen hair was increased as well. This suggested that the growth-phase in hair-growth-cycle (growth -> regression -> telogen -> growth) was extended by sufficient and prolonged delivery of hair growth ingredients to the dermal papilla cell.

Conclusion

PLGA NS could be utilized to create innovative medical or healthcare products with high safety and functional DDS performances toward next generation. As described, we have commercialized new DDS types of skincare cosmetics and hair growth tonics. Also, several pharmaceutical preparations and medical devices are now under clinical trials. Such the achievements have been evaluated as dozen prizes such as The Excellence Award of 2nd Monodukuri (Creation or Production) Japan Grand Prix (2007) and The 11th NAKAI Award of the Japan Society of Pharmaceutical Machinery and Engineering (2011) from related Ministries or Academic Societies in Japan. Further contributions of PLGA NS-based DDS to the medical, health and beauty care in the future are highly expected.

References

[1] Ishida, T. & Kiwada, H. (2004). Accelerated blood clearance of PEGylated liposomes after repeated injection. *Drug Delivery System*, *19*, 495-510.

[2] Kataoka, K., Harada, A. & Nagasaki, Y. (2001). Block copolymer micelles for drug delivery: design, characterization and biological significance. *Advanced Drug Delivery Reviews*, *47*, 113-131.

[3] Igarashi, R., Takenaga, M., Takeuchi, J., Kitagawa, A., Matsumoto, K. & Mizushima, Y. (2001). Marked hypotensive and blood flow-increasing effects of a new lipo-PGE1 (lipo-AS013) due to vascular wall targeting. *J. Control. Rel.*, *71*, 157-164.

[4] Kawashima, Y., Yamamoto, H., Takeuchi, H., Hino, T. & Niwa, T. (1998). Properties of a Peptide Containing DL-lactide/glycolide Copolymer Nanospheres Prepared by Novel Emulsion Solvent Diffusion Methods. Eurp. *J. of Pharm. Biopharma.*, *45*, 41-48.

[5] Murakami, H., Kobayashi, M., Takeuchi, H. & Kawashima, Y. (1999). Preparation of PLGA nanoparticles by modified spontaneous emulsification solvent diffusion method. *Int. J. Pharm.*, *187*, 143-152.

[6] Murakami, H., Kobayashi, M., Takeuchi, H. & Kawashima, Y. (2000). Further application of a modified spontaneous emulsification solvent diffusion method to various types of PLGA and PLA polymers for preparation of nanoparticles. *Powder Technol.*, *107*, 137-143.

[7] Kawashima, Y. (2006). Design of Poly(Lactic-co-glycolic Acid) (PLGA) Nanosphere for Developing to DDS. *J. Pharm. Sci. Technol.*, Jpn., *66*, 224-238.

[8] Toguchi, H., Ogawa, Y., Okada, H. & Yamamoto, M. (1991). Once-a-Month Injectable Microcapsules of Leuprorelin Acetate. *YAKUGAKU ZASSHI*, *111*, 397-409.

[9] Okada, H., Doken, Y., Ogawa, Y. & Toguchi, H. (1994). Preparation of three-month depot injectable microspheres of leuprorelin acetate using biodegradable polymers. *Pharm. Res.*, *11*, 1143-1147.

[10] Kawamura, K., (2006). GMP and Development of Pharmaceuticals in Japan from View Point of an Industry Scientist. *PDA Journal of GMP and Validation in Japan*, *8*, 2-17.

[11] Tahara, K., Sakai, T., Yamamoto, H., Takeuchi, H. & Kawashima, Y. (2008). Establishing chitosan coated PLGA nanosphere platform loaded with wide variety of nucleic acid by complexation with cationic compound for gene delivery. *International Journal of Pharmaceutics*, *354*(1), 210-216.

[12] Tahara, K., Sakai, T., Yamamoto, H., Takeuchi, H., Hirashima, N. & Kawashima, Y. (2009). Improved cellular uptake of chitosan-modified PLGA nanospheres by A549 cells. *International Journal of Pharmaceutics*, *382*(1), 198-204.

[13] Tahara, K., Yamamoto, H. & Kawashima, Y. (2010). Cellular uptake mechanisms and intracellular distributions of polysorbate 80-modified poly(D,L-lactide-co-glycolide) nanospheres for gene delivery. *European Journal of Pharmaceutics and Biopharmaceutics*, *75*(2), 218-224.

[14] Gwak, S. J. & Kim, B. S. (2008). Poly (lactic-co-glycolic acid) nanosphere as a vehicle for gene delivery to human cord blood-derived mesenchymal stem cells: comparison with polyethylenimine. *Biotechnology letters*, *30*(7), 1177-1182.

[15] Tahara, K., Samura, S., Tsuji, K., Yamamoto, H., Tsukada. Y., Bando, Y., Tsujimoto, H., Morishita, R. & Kawashima, Y. (2011). Oral nuclear factor-κB decoy oligonucleotides delivery system with chitosan modified poly(D,L-lactide-co-glycolide) nanospheres for inflammatory bowel disease. *Biomaterial*, *32*, 870-878.

[16] Fessi, H., Puisieux, F., Devissaguet, J. P., Ammoury, N. & Benita, S. (1989). Nanocapsule formation by interfacial polymer deposition following solvent displacement. *Int. J. Pharm.*, *55*(1), R1-R4.

[17] Bodmeier, R. & Cohen, H., 1990. Indomethacin polymeric nanosuspension prepared by microfluidization. *J. Control. Rel.*, *12*, 223-233.

[18] Allémann, E., Gurny, R. & Doelker, E. (1992). Preparation of aqueous polymeric nanodispersions by a reversible salting-out process, influence of process parameters on particle size. *Int. J. Pharm.*, *87*, 247-253.

[19] Kawashima, Y., Okumura, M. & Takenaka, H. (1982). Spherical crystallization: direct spherical agglomeration of salicylic acid crystals during crystallization. *Science*, *216*(4550), 1127-1128.

[20] Tsujimoto, H., Tsukada, Y., Hara, K. & Kawashima, Y. (2007). New particle design and its preparation of NFkB Decoy Oligodeoxynucleotide (NDON) loaded PLGA nanospheres for gene delivery. *J. Soc. Powder Technol., Japan*, *44*, 453-458.

[21] Tsukada, Y., Hara, K., Bando, Y., Huang, C. C., Kousaka, Y., Kawashima, Y. & Tsujimoto, H. (2009). Particle size control of poly (D,L-lactide-co-glycolide) nanospheres for sterile applications. *Int. J. Pharm.*, *370*(1), 196-201.

[22] Yamamoto, H., Kurashima, H., Katagiri, D., Yang, M., Takeuchi, H., Kawashima, Y., Yokoyama, T. & Tsujimoto, H., 2004, Poly (lactic-co-glycolic acid) nanosphere composite prepared with Mechanofusion® dry powder composition system for improving pulmonary insulin delivery with dry powder inhalation. *Journal of Pharmaceutical Science and Technology, Japan*, *64*(4), 245-253

[23] Yamamoto, H., Hoshina, W., Kurashima, H., Takeuchi, H., Kawashima, Y., Yokoyama, T. & Tsujimoto, H. (2004). Engineering of poly(D,L -lactic- co -glycolic acid) nanocomposite particles for dry powder inhalation dosage forms of insulin with the spray-fluidized bed granulating system. *J. Soc. Powder Technol., Japan*, *41*, 514-521

[24] Tsujimoto, H., Hara, K. & Kawashima, Y., (2005). Evaluation of glycaemia control in beagle dogs by the administration of insulin encapsulated PLGA nano-composite preparations. *J. Soc. Powder Technol., Japan*, *42*, 765-772

[25] Tsujimoto, H., Hara, K., Tsukada, Y., Kawashima, Y. & Hatano, S. (2007). Use of spouted bed type binderless granulation to design PLGA nano-composite granules for dry powder inhalation. *J. Soc. Powder Technol.*, Japan, *44*, 459-464

[26] Carino, G. P., Jacob, J. S. & Mathiowitz, E. (2000). Nanosphere based oral insulin delivery. *J. Control Rel.*, *65*, 261-269

[27] Kawashima, Y., Yamamoto, H., Takeuchi, H. & Kuno, Y. (2000). Mucoadhesive DL-lactide/glycolide copolymer nanospheres coated with chitosan to improve oral delivery of elcatonin. *Pharmaceutical Development and Technology*, *5*(1), 77-85.

[28] Lamprecht, A., Ubrich, N., Yamamoto, H., Schäfer, U., Takeuchi, H., Maincent, P., Kawashima, Y. & Lehr, C. (2001). Biodegradable nanoparticles for targeted drug delivery in treatment of inflammatory bowel disease. *J. Pharmacol. Exp. Ther.*, *299*, 775-781.

[29] Hara, K., Tsujimoto, H., Huang, C. C., Kawashima, Y. & Tsutsumi, M. (2008). Histological examination of PLGA nanospheres for intratracheal drug administration. *Int J Pharm*, *356*(1-2), 267-73.

[30] Hara, K., Tsujimoto, H., Huang, C. C., Kawashima, Y., Ando, R., Kusuoka, O., Tamura, K. & Tsutsumi, M. (2012). Ultrastructural and immunohistochemical studies on uptake and distribution of FITC-conjugated PLGA nanoparticles administered intratracheally in rats. *J Toxicol Pathol.*, *25*, 19-26.

[31] Morishita, R., Sugimoto, T., Aoki, M., Kida, I., Tomita, N., Moriguchi, A. & Ogihara, T., 1997. In vivo transfection of cis element "decoy" against nuclear factor-κB binding site prevents myocardial infarction. *Nature medicine*, *3*(8), 894-899.

[32] Sawa, Y., Morishita, R., Suzuki, K., Kagisaki, K., Kaneda, Y., Maeda, K. & Matsuda, H., 1997. A novel strategy for myocardial protection using in vivo transfection of cis element'decoy'against NFkappaB binding site: evidence for a role of NFkappaB in ischemia-reperfusion injury. *Circulation*, *96*(9 Suppl), II-280.

[33] Tsukada, Y., Tsujimoto, H., Hara, K., Sakaguchi, M., Aoki, M., Morishita, R. & Kawashima, Y. (2006). The developments of NFκB Decoy Oligodeoxynucleotides loaded PLGA nanosphere and the applications for atopic dermatitis. 2nd International *Technical Forum Inspiring Powder Technology*, 45-46, November 9, Japan.

[34] Tsukada, Y., Tsujimoto, H., Watanabe, H., Sugimoto, T., Nakano, O., Makino, H., Yamamoto, H. & Morishita, R. (2013). Development of oral formulation technology for nucleic acid drug by using PLGA nanoparticles as DDS carriers. *J. Soc. Powder Technol.*, Japan, *50*, 513-518.

[35] Shim, J. (2007). Transdermal applications of nanoparticulates. *DRUGS AND THE PHARMACEUTICAL SCIENCES*, *166*, 327.

[36] Chen, D. L., Zheng, D. & Paller, A. S. (2013). Nano-based gene therapy for dermatologic diseases. In Nanotechnology in Dermatology (109-117). Springer New York.

[37] Da Silva, C. L., Del Ciampo, J. O., Rossetti, F. C., Bentley, M. V. & Pierre, M. B., (2013). Improved in vitro and in vivo cutaneous delivery of protoporphyrin IX from PLGA-based nanoparticles. *Photochemistry and photobiology*, *89*(5), 1176-1184.

[38] Tsujimoto, H., Hara, K., Huang, C. C., Yokoyama, T., Yamamoto, H., Takeuchi, H., Kawashima, Y., Akagi, K. & Miwa, N. (2004). Percutaneous absorption study of biodegradable PLGA nano-spheres via human skin biopsies. *J. Soc. Powder Technol.*, Japan, *41*, 867-875.

[39] Tsukada, Y., Sasai, A, Tsujimoto, H., Yamamoto, H., Kawashima, Y. & Miwa N., 2013. Practical use of PLGA nanospheres used as DDS carriers for cosmetics. *COSMETIC STAGE*, *7*(4), 42-51.

[40] Virmani, R., Guagliumi, G., Farb, A., Musumeci, G., Grieco, N., Motta, T. & Kolodgie, F. D., 2004. Localized hypersensitivity and late coronary thrombosis secondary to a sirolimus-eluting stent should we be cautious?. *Circulation*, *109*(6), 701-705.

[41] Joner, M., Finn, A. V., Farb, A., Mont, E. K., Kolodgie, F. D., Ladich, E. & Virmani, R., 2006. Pathology of drug-eluting stents in humans: delayed healing and late thrombotic risk. *Journal of the American College of Cardiology*, *48*(1), 193-202.

[42] Nordmann, A. J., Briel, M. & Bucher, H. C. (2006). Mortality in randomized controlled trials comparing drug-eluting vs. bare metal stents in coronary artery disease: a meta-analysis. *European Heart Journal*, *27*(23), 2784-2814.

[43] Nakano, K., Egashira, K., Masuda, S., Funakoshi, K., Zhao, G., Kimura, S., Matoba, T., Sueishi, K., Endo, Y., Kawashima, Y., Hara, K., Tsujimoto, H., Tominaga, R. & Sunagawa, K. (2009). Formulation of nanoparticle-eluting stents by a cationic electrodeposition coating technology. JACC: *Cardiovascular Interventions*, *2*(4), 277-283.

[44] Tsukada, Y., Hara, K., Bando, Y. & Tsujimoto, H. (2009). Development of PLGA nanoparticles-coated stents. *FUNTAI GIJUTSU*, *1*(12), 53-59.

[45] Tsukada, Y., Tsujimoto, H., Yamamoto, H. & Kawashima, Y. (2012). Development of drug-eluting stent coated with PLGA nanoparticles. *Chemical Industry*, *63*(7), 52-58.
[46] AnGes, MG, Inc., Medikit Co. Ltd. and Hosokawa Micron Corporation, Press Release, http://www.anges-mg.com/news/pdf/081104.pdf, November 4, 2008
[47] AnGes MG, Inc., Medikit Co. Ltd. and Hosokawa Micron Corporation, Press Release, http://www.anges-mg.com/news/pdf/20120328.pdf, March 28, 2012
[48] Yoshimura, S., Morishita, R., Hayashi, K., Yamamoto, K., Nakagami, H., Kaneda, Y. & Ogihara, T. (2001). Inhibition of intimal hyperplasia after balloon injury in rat carotid artery model using cis-element'decoy'of nuclear factor-kappaB binding site as a novel molecular strategy. *Gene Therapy*, *8*(21), 1635-1642.
[49] Takeuchi, K., Itoh, H., Yonemitsu, Y., Matsumoto, T., Kume, M., Komori, K. & Maehara, Y. (2007). In vivo reduction of the nuclear factor-κB activity using synthetic cis-element decoy oligonucleotides suppresses intimal hyperplasia in the injured carotid arteries in rabbits. *Surgery Today*, *37*(7), 575-583.
[50] Ohtani, K., Egashira, K., Nakano, K., Zhao, G., Funakoshi, K., Ihara, Y. & Sunagawa, K., (2006). Stent-based local delivery of nuclear factor-κB decoy attenuates in-stent restenosis in hypercholesterolemic rabbits. *Circulation*, *114*(25), 2773-2779.
[51] Miyake, T., Aoki, M., Shiraya, S., Tanemoto, K., Ogihara, T., Kaneda, Y. & Morishita, R. (2006). Inhibitory effects of NFκB decoy oligodeoxynucleotides on neointimal hyperplasia in a rabbit vein graft model. *Journal of Molecular and Cellular Cardiology*, *41*(3), 431-440.
[52] Yamasaki, K., Asai, T., Shimizu, M., Aoki, M., Hashiya, N., Sakonjo, H. & Morishita, R. (2003). Inhibition of NFκB activation using cis-element 'decoy'of NFκB binding site reduces neointimal formation in porcine balloon-injured coronary artery model. *Gene Therapy*, *10*(4), 356-364.
[53] Shimizu, N., Azuma, N., Nishikawa, T., Hirata, S., Morishita, R., Kaneda, Y. & Sasajima, T. (2007). Effect on vein graft intimal hyperplasia of nuclear factor-kB decoy transfection using the second generation of HVJ vector. *The Journal of Cardiovascular Surgery*, *48*(4), 463-470.
[54] Miyake, T., Ihara, S., Nakagami, H. & Morishita, R. (2012). Prevention of neointimal formation after angioplasty using Nf-Kappab decoy oligodeoxynucleotide-coated balloon catheter in rabbit model. *Journal of Hypertension*, *30*, e277-e278.
[55] Tsujimoto, H. (2006). Development of functional skin and scalp care cosmetics using biodegradable PLGA nanospheres. *Drug Delivery System*, 21-4, 405-416.
[56] Elias, P. M. (1983). Bpidermal lipids. barrier function, and desquamalion. *J. Invest. Dermatol.*, *80*, 44-49.
[57] Sasai, A., Matsuzaki, K., Tsujimoto, H., Yamamoto, H., Kawashima, Y. & Miwa, N. (2013). PLGA nano-sphere used as DDS carriers for hair restorer. *Fragrance Journal*, *41*(11), 49-55.
[58] Tsujimoto, H., Hara, K., Tsukada, Y., Huang, C. C., Kawashima, Y., Arakaki, M., Okayasu, H., Mimura, H. & Miwa, N. (2007). Evaluation of the permeability of hair growing ingredient encapsulated PLGA nanospheres to hair follicles and their hair growing effects. *Bioorganic & medicinal chemistry letters*, *17*, 4771-4777.
[59] Tsujimoto, H., Hara, K., Bando, Y. & Tsukada, Y. (2007). Evaluation of the permeability of hair growing ingredient encapsulated PLGA nanospheres to hair follicles and their hair growing effects. *Fragrance Journal*, *12*, 55-62.

In: New Developments in Polylactic Acid Research
Editor: Courtney Winthrop
ISBN: 978-1-63463-054-2

Chapter 7

Reactive Compatibilization of PLA-TPS Blends

M. H. Gutiérrez Villarreal[1,•]***, F. J. Rodríguez González***[1,*]
and M. C. Kantun Uicab[1]
[1]Centro de Investigación en Química Aplicada, Saltillo, Coahuila, México

Abstract

Reactive compatibilization of polylactic acid (PLA)/thermoplastic starch (TPS) blends using methyl methacrylate-glycidyl methacrylate (GMA-MMA) copolymer as compatibilizer was reported in this chapter. During melt blending, the reduction of torque of the uncompatibilized PLA/TPS blends was related to the breakup of starch chains. Moreover, the chemical reaction between MMA-GMA copolymer with PLA and TPS resulted, on one hand, the chain extension of PLA and, on the other hand, torque equilibrium due to the competition among glycerol-MMA-GMA reaction, starch-chain breakup and the formation of the actual compatibilizer MMA-GMA-g-starch. The effect of melt blending time and the concentration of MMA-GMA copolymer on the compatibilization reaction was studied. As the reaction time increased the formation of the actual compatibilizer produced a drop of the TPS domains. On the other hand, the variation of MMA-GMA copolymer concentration evidenced a competition between PLA chain extension and the formation of the actual compatibilizer MMA-GMA-g-starch. Reactive compatibilization of PLA/TPS blends resulted in ductile materials which could have potential application in packaging and agricultural mulch.

• E-mail: mario.gutierrez@ciqa.edu.mx.
* E-mail: francisco.rodriguez@ciqa.mx.

Introduction

PLA is one of the most important biodegradable polyesters with wide applicability in many fields, such as tissue engineering, drug delivery systems, biocompatible/bioabsorbable medical device market and packaging industry. Recently, PLA has been considered as a major alternative to petroleum-based plastics for disposable items, such as trash bags, containers, drinking cups, overwrap, lamination films, blister packages, and food utensils. PLA has a good biocompatibility, thermoplasticity, processability, and good physical properties, such as high strength and high modulus. It can be processed by specific manufacturing processes, such as injection molding, sheet extrusion, blow molding, thermoforming, blown film, and fiber spinning, which may be carried out using conventional techniques and processing equipment. However, PLA is very brittle under tensile and bending loads and will develop serious physical aging during application. Moreover, PLA is a much more expensive material than the common industrial polymers and its applications are somewhat limited due to its high cost (3-4 euros /Kg). PLA consumption in 2007 was about 60 thousand tons with a growing demand per year of 10-15 %, it is expected for 2015 a consumption of 140-150 thousand tons (Polylactic acid (PLA)). The consumption of commercial bioplastic products is being hampered by the higher price and shortage of raw vegetable, for this reason researchers around the world are working on the reduction of its cost by blending PLA with starch.

Thermoplastic Starch (TPS)

Starch is a renewable, degradable, carbohydrate biopolymer derived from abundant and readily available sources. Native starch, for industrial application, is about $0.09/lb. Starch-based materials have been considered as a potential sustainable replacement for oil-based synthetic polymers. Starch is a natural hydrophilic polymer which can be extracted from plants in granular form. It can be transformed into a thermoplastic material using the same equipment for the extrusion of common plastics. Figure 1 shows a schematic representation of the screw configuration of the twin-screw extruder used by the preparation of thermoplastic starch (TPS) (Sandoval, 2012).

It has been extensively reported in the literature that rheological, thermal and mechanical properties of TPS are dependent of glycerol content (Rodriguez-Gonzalez 2004). Figure 2 shows the apparent shear viscosity, measured at 180°C in a capillary rheometer, of TPS materials plasticized with 29, 36 and 40% glycerol.

All TPS showed a shear-thinning behavior in the whole range of evaluation (Rodriguez-Gonzalez 2004). As expected, shear viscosity decreased as glycerol content increased. For instance, apparent shear viscosity of TPS29 drops 28% at constant shear speed rate when glycerol content increased to 40%.

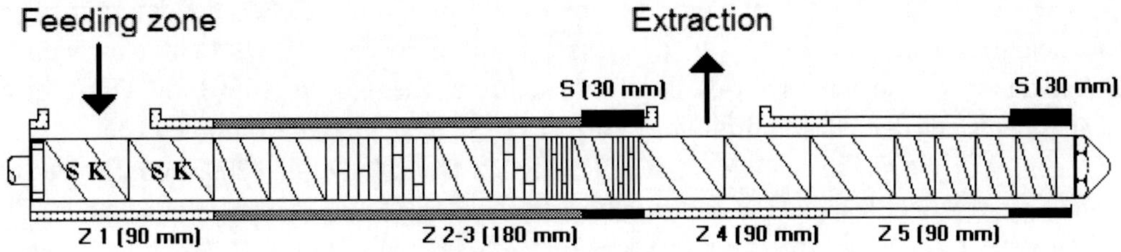

Figure 1. Configuration of twin screw extruder ZSK30 for the preparation of ATP.

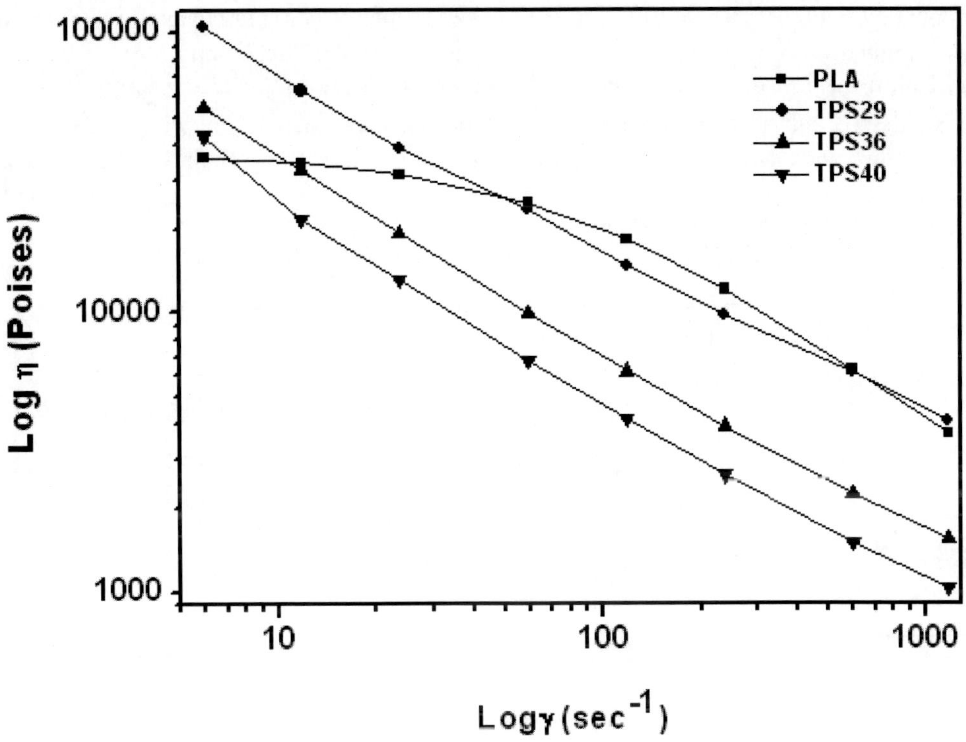

Figure 2. Apparent shear viscosity of PLA and TPS materials plasticized with 29, 36 and 40% glycerol at 180°C in a capillary rheometer.

In spite of the commercial and ecological advantages of starch, starch-based materials present two main drawbacks: brittleness (Stenhouse et al., 1993, Lawton and Fanta, 1994, Bikiaris et al., 1998, Averous et al., 2001) and moisture sensitivity (Loomis et al., 1993, Shogren, 1993, Stenhouse et al., 1993, Van Soest and Borger, 1997), which can be minimized by melt blending with synthetic polymers.

PLA/TPS Blends

Melt blending with immiscible polymers has been considered as an excellent approach to the development of new plastic materials (Utracki, 1989). Multi-phase systems are, in general, characterized by a dispersed component embedded in a continuous matrix. Mechanical performance of multi-phase systems is closely related to the size and shape of the dispersed domains. The morphology of immiscible polymer blends is determined by rheological (viscosity, elasticity, yield stresses and microscopic flow fields) as well as thermodynamic parameters, such as interfacial tension (Elemndorp, 1986). St. Pierre et al. (1997) suggested that the morphology of PE/TPS blends can controlled due to the deformability of the starch domains. Later, Rodriguez-Gonzalez et al. (2003) confirmed that morphology of PE/TPS blends is dependent of the viscosity ratio of the polymer pairs. The apparent shear viscosity of different TPS and PLA is depicted in Figure 2. It can be observed that the viscosity ratio of all PLA/TPS pairs is lower than the unity at 100 s^{-1}, which is considered the shear rate for extrusion process. PLA was melt blended with 20% of TPS (compounded with 29, 36 and 40% glycerol) using an internal mixing chamber at 180°C and 60 rpm (Figure 3). The addition of 20% TPS resulted in gradual reduction of torque values at 6 min with respect to PLA. This reduction seemed to be proportional to the glycerol content in TPS. After 15 min of mixing, torque of PLA dropped less than 1 Nm, while the torque of all PLA/TPS blends decreased more than 2 Nm, probably due to the chain breakup of starch molecules and the plasticizing role of glycerol at the interfase of PLA/TPS domains (Rodriguez-Gonzalez 2003, Tauget 2009).

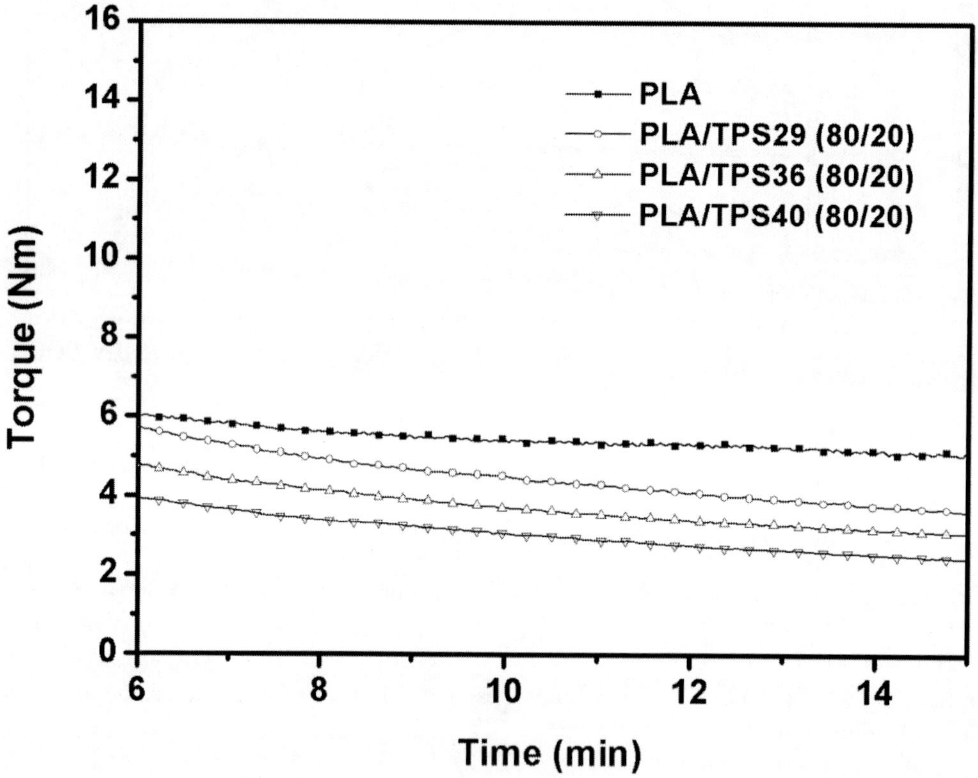

Figure 3. Rheograms of PLA and PLA blended with 20% of TPS (having 29, 36 and 40% glycerol) at 180°C and 60 rpm.

After melt blending, samples were microtomed at room temperature and the TPS domains were extracted by acid hydrolysis. The morphology of PLA/TPS blends melt blended in the internal mixer showed that the particle size was unexpectedly similar regardless the very different viscosity showed by TPS compounded with 29, 36 and 40%

glycerol (Figure 4). It has been reported that the reduction of the viscosity of TPS dispersed phase resulted in finer particle size (Rodriguez-Gonzalez 2003, Tena 2008, Tauget 2009). However, Mendez et al. (2009) and Li et al. (2011) observed that at viscosity ratios between 0.29 and 0.89 there was not differences in the morphology of PE/TPS and PCL/TPS blends, respectively.

The evaluation of tensile properties of PLA and PLA/TPS blends is depicted in Figure 5. It was observed that PLA showed a fragile behavior. As expected, the addition of 20% of TPS does not change the fragile behavior, but also resulted in an important drop of both the Young´s Modulus and the maximum tensile strength.

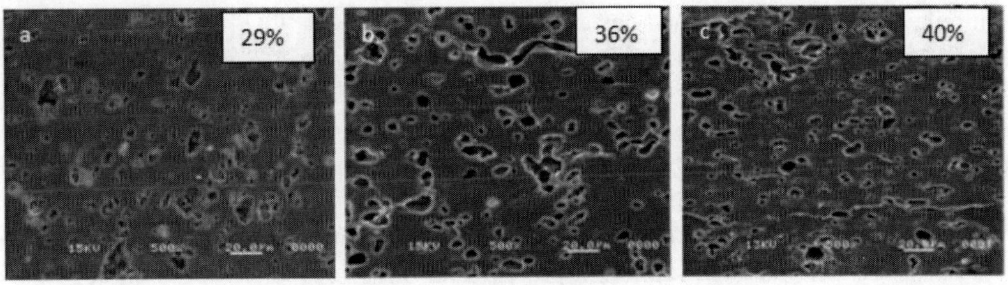

Figure 4. Morphology of PLA blended with TPS compounded with 29, 36 and 40% glycerol.

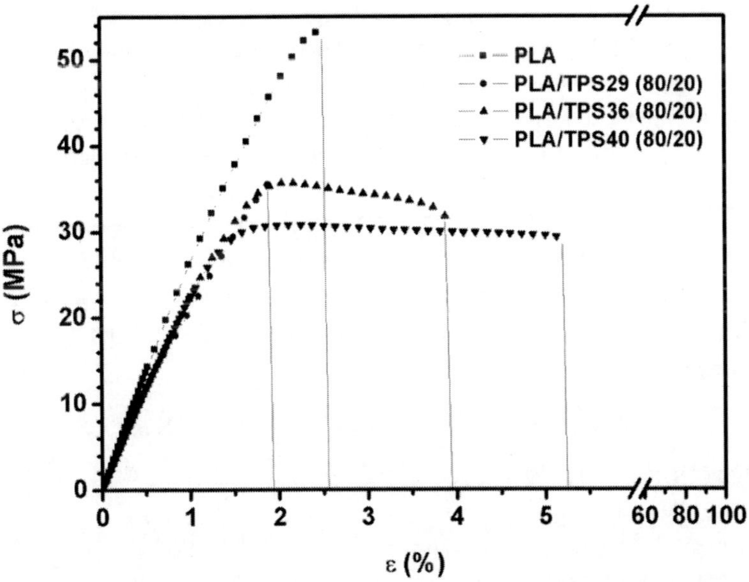

Figure 5. Stress-strain curves of PLA and its blends with TPS at different glycerol content.

The drop in stiffness was expected because the maximum tensile strength of TPS has been reported to be between 1 to 34 MPa when the glycerol content varied from 39 to 33%, respectively (Sandoval 2014).

Reactive Compatibilization of PLA/TPS Blends

Melt blending of PLA and TPS resulted in weak mechanical properties, such as the maximum tensile strength and Young´s Modulus, due to the high interfacial tension between the hydrophilic TPS and the hydrophobic PLA (Wang 2008, Averous 2000). In order to reduce the interfacial tension between PLA and TPS, it is necessary to use a proper interfacial agent. The most common alternative to reduce the interfacial tension of immiscible polymer systems is the addition of a block copolymer having segments that interact with each one of blend components. Interfacial agents, such as SBS block copolymer, which is a typical example of compatibilizer for the PE-PS system where the styrene blocks present on the SBS are miscible with the PS phase and butadiene blocks interact with the PE. Another alternative for development of interfacial agents is the use of copolymers in which one of the comonomers promote miscibility with one of the phases and the other present strong interactions or chemical reactions with the other phase. An example of these copolymers could be a LLDPE grafted with MAH, in which the MAH could react with the amine end-group of polyamide or the hydroxyl end-group of polyester.

In 2006, Li and Huneault reported the development of a process of grafting MAH on PLA, which was tried as interfacial modifier in blends of PLA/TPS. It was observed that the addition of PLA-g-MAH decreases dramatically the TPS domains due to the decrease of interfacial tension at the PLA/TPS interface. Moreover, the presence of the compatibilizer improved dramatically the ductility of the blends, but it does not affect stress resistance and Young’s modulus. A probable disadvantage of the process for obtaining the interfacial agent is that the viscosity of PLA-g-MAH could be difficult to control because it decreased more than one order of magnitude with the addition of only 0.5% of peroxide.

Recently, it was proposed by our research group to study the influence of the architecture of acrylic copolymers on the compatibilization efficiency of PLA/TPS blends (Kantun-Uicab, 2014). Copolymers of methyl methacrylate (MMA)-glycidyl methacrylate (GMA) having around 30, 60 or 90 kDa and different MMA-GMA ratios were synthesized and evaluated as compatibilizers for PLA/TPS blends. In order to elucidate how the acrylic copolymer could react with PLA end-groups and TPS hydroxyl groups during reactive compatibilization, PLA/MMA-GMA and TPS/MMA-GMA (80:20) blends were melt blended in a Brabender internal mixer (Figure 6).

Chemical Reactions Between Epoxide Groups And PLA And TPS

The addition of MMA-GMA copolymer to PLA resulted in a progressive increment of torque during the whole mixing period. This torque increasing is an evidence of a viscosity increase and it can be explained by the reactions between the epoxy groups present on the GMA and the hydroxyl and carboxylic terminal groups of PLA. This torque behavior has been observed on other similar studies with reactive copolymers (Liu et al. 2011). In Figure 3, it was observed that the torque decreased during melt blending PLA and TPS probably due to the breakup of starch molecules. The addition of 20% MMA-GMA copolymer to TPS40, unexpectedly, resulted in a constant torque value during the whole melt-mixing process. It could be explained, on one hand, by the reaction of hydroxyl groups of glycerol with epoxy groups of GMA. It has been reported that a thin layer of glycerol could be formed at the interfase of TPS polymer blends (Rodriguez-Gonzalez 2003, Tauget 2009) and it could react with the reactive copolymer. On the other hand, it seemed that an equilibrium between chain breakup and chain extension of starch molecules could be reached with the remaining epoxy groups. These results suggested that very different behavior of torque during melt blending of PLA/TPS depending on if epoxy-groups react with PLA end-groups of TPS hydroxyl groups.

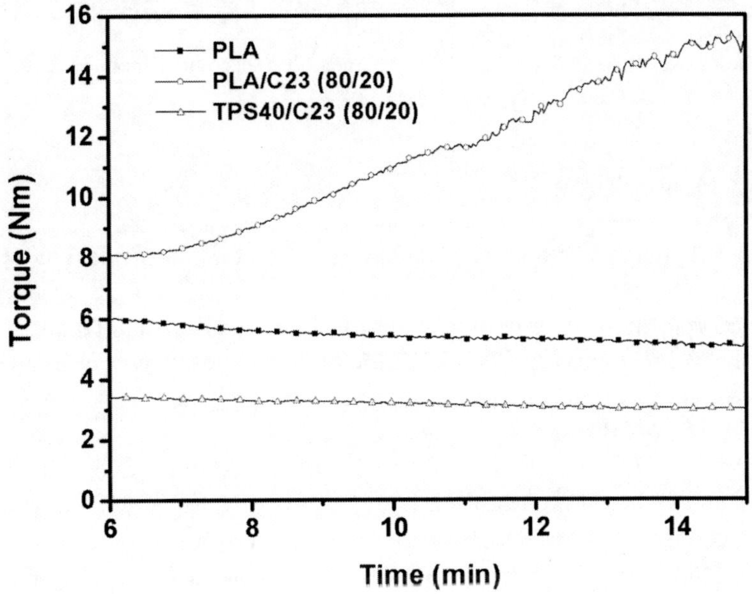

Figure 6. Rheograms of PLA, PLA/MMA-GMA (80:20) and TPS40/MMA-GMA copolymer (80:20).

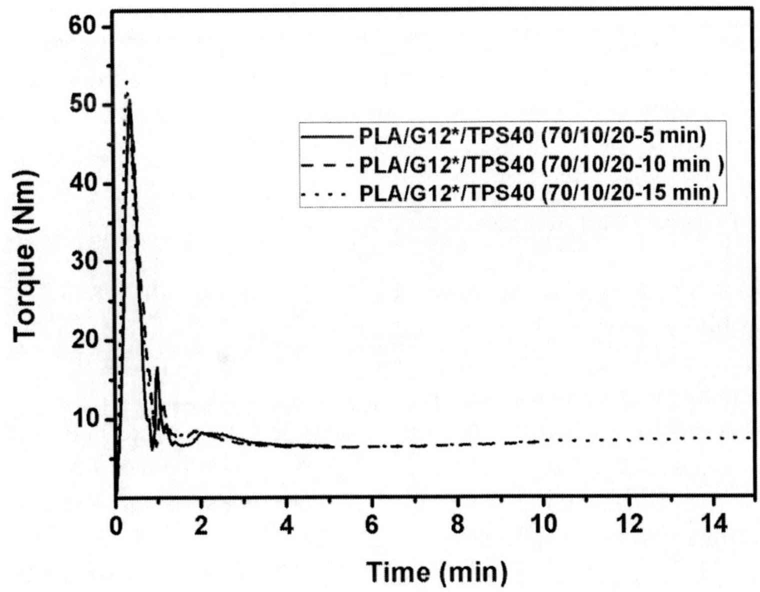

Figure 7. Torque-time graphic of PLA/MMA-GMA/TPS40 blends (70:10:20) at 180°C and 60 during 5, 10 and 15 min.

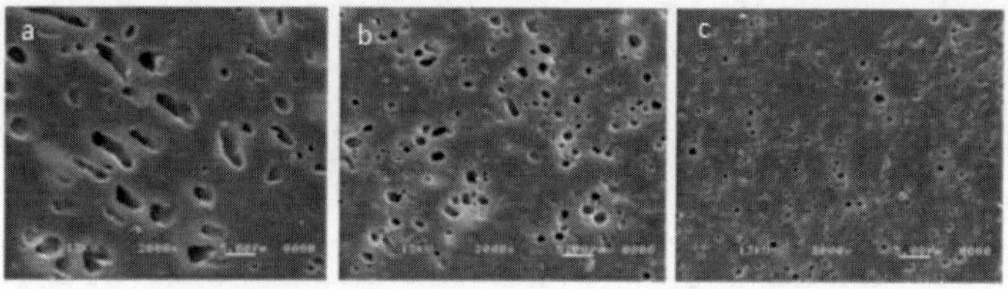

Figure 8. Morphology of PLA/MMA-GMA/TPS40 blends (70:10:20) melt-blended during 5, 10 and 15 min.

The Effect of Mixing Time

As any other chemical reaction, reactive compatibilization is dependent of mixing time as observed in Figures 7 and 8. PLA/MMA-GMA/TPS40 blends (70:10:20) were melt blended at 180°C and 60 rpm in an internal mixer during 5, 10 and 15 min. Torque-time graphics demonstrated that the melt mixing process is quite reproducible (Figure 7). On the other hand, Figure 8 shows how the reaction time affected the *in situ* formation of the actual compatibilized for PLA/TPS blends.

As mentioned above, epoxy groups of MMA-GMA copolymer could react with carboxylic acid and hydroxyl groups of PLA, glycerol or starch, but the formation of MMA-GMA-g-starch copolymer gives place to the actual compatibilizer. After 5 min of melt mixing, there are many big and elongated TPS40 particles that seemed to be larger than 5 μm and only a few small ones (Figure 8a), however, this morphology is much finer than the one observed for PLA/TPS40 (Figure 4c) which evidenced that interfacial tension decreased due to the *in situ* formation of MMA-GMA-g-starch copolymers. The evolution of interfacial tension was evidenced at 10 and 15 min of mixing (Figures 8b and 8c, respectively), where the number of the TPS40 domains smaller than 1 μm of increased as a function of time. It is important to highlight that the distance between TPS particles decreased as the mixing time increased, which is an indication of better dispersion.

The Effect of Copolymer Concentration

The concentration of reactive copolymer, such as the reaction time, is a key parameter for reactive compatibilization as observed in Figures 9 and 10. PLA/TPS36 blends were compatibilized in an internal mixer at 180°C and 60 rpm using 0, 3, 5 and 7% of MMA-GMA copolymer and the torque versus mixing time is depicted in Figure 9. As previously mentioned, when no MMA-GMA copolymer was added the torque of PLA/TPS36 blend decreased and the breakup of starch molecules was evidenced. When 3 and 5% of MMA-GMA copolymer was included in the formulation, the torque at 6 min of mixing was about 1.7 Nm higher than the uncompatibilized blend, which was an indication that PLA chain extension occurred even at short mixing times. At 15 min., PLA chain extension produced an important increment of torque of PLA/TPS36 blend which was about 2 and 3 times higher for blends compounded with 3 and 5% MMA-GMA copolymer, respectively, with respect to the uncompatibilized blend.

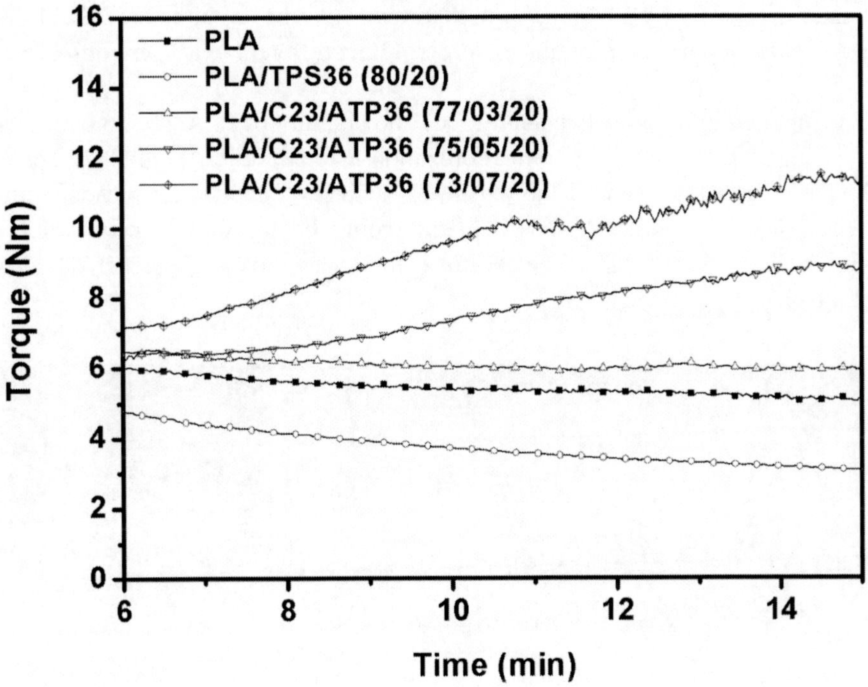

Figure 9. Rheogram of PLA/TPS36 blends compatibilized with 0, 3, 5 and 7% of MMA-GMA copolymer at 180°C and 60 rpm.

When PLA/TPS36 blends were formulated with 7% MMA-GMA copolymer, PLA chain extension was even more dramatic. Torque improved from 4.8 to 7.1 Nm and from 3 to 11 Nm at 6 and 15 min., respectively. These results evidenced that the chain extension reaction of PLA increased as both the concentration of MMA-GMA copolymer and reaction time increased. Chemical reaction between starch and MMA-GMA was not evident in torque vs time graphics. However, the analysis of the evolution of TPS particle size as a function of MMA-GMA copolymer content could give indications about the *in situ* formation of the actual compatibilizer, i.e. MMA-GMA-g-starch copolymer.

The effect of the concentration of MMA-GMA (1, 5 and 10%) copolymer on the morphology of PLA/TPS blends compatibilized in an internal mixer at 180°C and 60 rpm is presented in Figure 10.

As observed in Figure 4b, the uncompatibilized PLA/TPS36 blends showed some elongated TPS36 particles larger than 20 μm and some a few particles below 5 μm. The addition of only 1% MMA-GMA copolymer produced an important reduction of the interfacial tension resulting in TPS36 particles where the largest TPS36 particles were below 5 μm and there were also a few smaller than 1 μm.

These results were indicative than just a small concentration of the reactive copolymer is enough to produce the actual compatibilizer MMA-g-starch, even though the epoxy groups could also react with PLA end-groups and glycerol hydroxyl groups. At higher MMA-GMA concentration, it was not so evident the reduction of the size of TPS36 particles, which is indicative that the interface between PLA and TPS36 is being saturated by the actual compatibilizer. The results discussed above evidenced that there exist a competition between chain extension reactions of PLA molecules and the in situ formation of the actual compatibilizer MMA-GMA-g-starch, and that tailoring the architecture of MMA-GMA copolymers could be the key for the control of the different reactions observed in this complex system.

Such as reported by Li and Huneault (2006), the ductility of PLA/TPS36 blends could be dramatically improved by reactive compatibilization as observed in Figure 11. The addition of MMA-GMA copolymer shifted the tensile behavior of PLA/TPS36 blends from fragile (Figure 5) to ductile. Ductile PLA/TPS blends could be useful for the development of applications such as plastic bags, mulch films and many other where the ductility is an important factor.

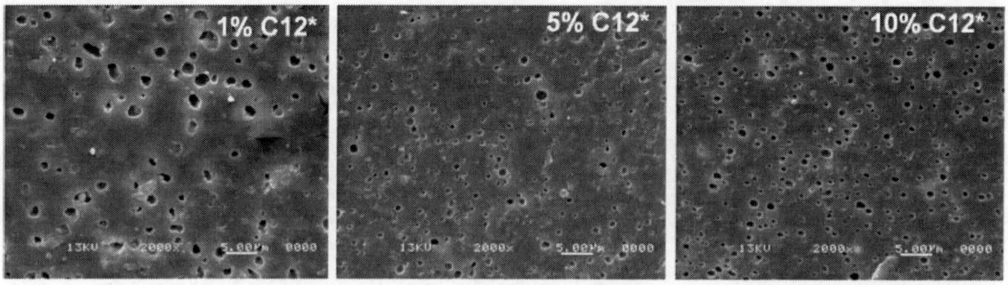

Figure 10. Morphology evolution of TPS particle size as a function of MMA-GMA copolymer content.

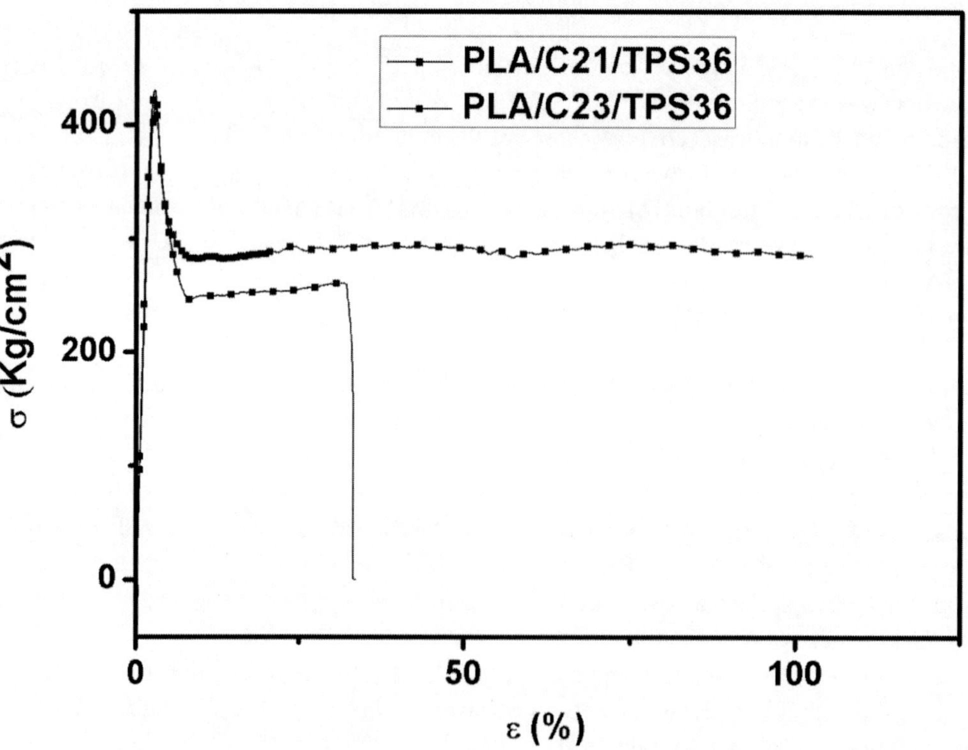

Figure 11. Stress-strain curves of PLA/TPS36 blends compatibilized with MMA-GMA copolymer.

Conclusion

PLA/TPS blends are presented as new plastic materials which can potentially substitute oil-based materials. However, PLA/TPS blends have poor mechanical performance due to the high interfacial tension between the hydrophobic PLA and hydrophilic TPS. In order to reduce the interfacial tension and, consequently, improve the mechanical performance an acrylic copolymer was evaluated as a compatibilizer for PLA/TPS blends. The presence of the MMA-GMA copolymer in PLA/TPS blends evidenced that multiple chemical reactions can occur during the melt blending process. Hence, epoxy groups of MMA-GMA copolymer can react with the PLA end-groups giving place to the chain extension of the PLA matrix. When the MMA-GMA copolymer reaches the interface of PLA/TPS blends, some of the epoxy groups can be eliminated by the reaction with glycerol hydroxyl groups. The remaining epoxy groups can react with starch hydroxyl groups giving place to the actual in situ formed compatibilizer (MMA-GMA-g-starch). It is proposed that tailoring the molecular architecture of MMA-GMA copolymer would control the extent of each the chain extension or in situ formation of the actual compatibilizer reactions.

References

Averous, L., L. Moro, P. Dole, C. Fringant, *Polymer*, 41 (11), 4157-4167 (2000).

Bikiaris, D., J. Prinos, K. Koutsopoulos, N. Vouroutzis, E. Pavlidou, N. Frangis, C. Panayiotou, *Polymer Degradation and Stability*, 59, Issues 1–3, 287-291 (1998).

Elmendorp J. J. and A. K. Van Der Vegt, *Polymer Engineering and Science*, 26 (19), 1332–1338 (1986).

Ho, C.-H., C.-H. Wang, C.-I. Lin, Y.-D. Lee, *Polymer,* 49, 3902-3910 (2008).

Li, H., M. A. Huneault, *Journal of Applied Polymer Science*, 122, 134-141 (2011).

Kantún Uicab, M. C., Ph. D. Thesis, Centro de Investigación en Química Aplicada, Saltillo, México (2014).

Lawton, J. W. and G. F. Fanta., *J. Carbohydr. Polym.*, 23 :275-280. (1994).

Li G., P. Sarazin, W. J. Orts, S. H. Imam, B. D. Favis, *Macromolecular Chemistry and Physics,* 212 (2011) 1147-1154.

Liu H., Song W., Chen F., Guo L. Zhang L. *Macromolecules,* 2011, 44, 1513-1522.

Loomis G. L., Hopkins A. R., George E. R., Starch-based materials, Ching C., Kaplan D. L., Thomas E. L., Biodegradable polymers and packaging, 42 - 51, Technomic Publishing Co. Inc, Lancaster (1993).

Martin, O., L. Avérous, *Polymer,* 42 (2001) 6209-6219.

Méndez Hernandez, M. L., Ph. D. Thesis, Centro de Investigación en Química Aplicada, (2009).

Polylactic acid (PLA): properties, market and perspectives, http://biopol.free.fr/?p=282.

Rodríguez-González F. J., Ramsay B. A., Favis B. D. *Polymer,* 44, 1517-1526 (2003)

Rodriguez-Gonzalez F. J., B. A. Ramsay, B. D. Favis, *Carbohydrate Polymers*, 58 (2004) 139-147.

Sandoval Arellano, Z. Ph. D. Thesis, Centro de Investigación en Química Aplicada, Saltillo, México (2012).

Sandoval Arellano, Z., C. S. Tena Salcido, M. L. Méndez Hernández, J. C. Contreras Esquivel, M. H. Gutiérrez Villarreal, F. J. Rodríguez González, Revista Mexicana de Ingenieria Química (in press, 2014).

Shogren, R. L., G. F. Fanta, and W. M. Doane, *Starch/Starke*, 45, 276 (1993).

Stenhouse, P. J., J. M. Mayer, M. J. Hepfinger, E. A. Costa, D. L. Kaplan, C. Ching, D. L. Kaplan, E. Thomas (Eds.), Biodegradable Polymers and Packaging, Technomic, Lancaster, Basel, pp. 151–158 Chap. 11 (1993).

St-Pierre, N., B. D. Favis, B. A. Ramsay, J. A. Ramsay, and H.Verhoogt, *Polymer*, 38 (3), 647-655 (1997).

Taguet, A., M. A. Huneault, B. D. Favis, *Polymer*, 50 (2009) 5733-5743.

Tena-Salcido, C. S., F. J. Rodríguez-González, M. L. Méndez-Hernández, J. C. Contreras-Esquivel. *Polymer Bulletin*, 60, 677-688 (2008).

Utracki, L. A., Polymer Blends Handbook, Kluwer Academis Publishers, 2003.

van Soest, J. J. G. and D. B. Borger, *J. Appl. Polym. Sci.,* 64, 631-644 (1997).

Index

B

C

F

G

H

I

J

K

L

M

N

O

P

Q

R

S

T